GEOGRAPHIC TRAVELS IN CENTRAL AUSTRALIA

To Len Beadell

Who inspired me to produce this facsimile.

ISBN 0 646 24262 8

GEOGRAPHIC TRAVELS IN CENTRAL AUSTRALIA
FROM 1872 TO 1874

First published in Melbourne by M'Carron, Bird & Co.,
1875.

First facsimile edition published by Corkwood Press 1993.
Second facsimile edition published 1995.

Giles, Ernest. 1835-1897.
Geographic Travels in Central Australia
from 1872 to 1874.

Facsimile reprint. First published: Melbourne:
M'Carron, Bird & Co., 1875.
Ferguson 9912.

Printed in Australia for Corkwood Press in collaboration with Westprint.
26 Crofton Street
Bundaberg
Q 4670

GEOGRAPHIC TRAVELS

IN

CENTRAL AUSTRALIA

FROM 1872 TO 1874,

BY

ERNEST GILES,

KNIGHT OF THE ORDER OF THE CROWN OF ITALY,
HONORARY MEMBER OF THE ROYAL ITALIAN GEOGRAPHIC SOCIETY.

Melbourne:

PRINTED FOR THE AUTHOR BY M'CARRON, BIRD & CO.,
37 FLINDERS LANE WEST.

1875.

INTRODUCTION.

The able and undaunted traveller, whose diaries are re-issued in these pages, has entrusted to me the task of revising his journals for the press and introducing this volume by a few prefatory words to public notice. I have complied with his request all the more readily, as he has entered already anew the field of geographic exploration. It was deemed desirable to render the instructive and stirring account of his exploits accessible in a more attainable form than that of parliamentary records. No one perusing these journals will withhold the admiration due to the courage, skill and perseverance displayed by Mr. Giles under unusually trying difficulties, more particularly so as he was supported only by the most slender means; and it seems probable, that any other explorer would have shrunk from the task of braving the terrors of the Australian deserts in seasons of protracted aridity with such narrow resources as were at his commands.

The writer of these lines may possibly be held responsible for having unwarrantedly initiated and encouraged these enterprises, though now of lasting importance, with such utterly inadequate auxiliaries; but although aware that the greatest deeds in Australian geographic research have generally been accomplished with the most modest resources, yet he was equally cognisant that ampler appliances, if accessible, would lessen the dangers and privations which beset the path of his now deservedly celebrated friend. It is therefore due on this occasion to relate, under which circumstances these two expeditions arose.

The formation of our Trans-continental Telegraph-Line, a few years ago, gave facilities for carrying explorations from some central positions of our continent widely westward, and afforded also easier means of retreat in cases of distress. Moreover, a few previous seasons rather favourable for rain held out the hope of traversing Central Australia then with comparative ease; and in order that these chances might not be lost I induced my experienced friend, who for years was eager to push in new lines through the interior, to step forward as the champion of a new cause of science. I long had entertained the view, that the remotest inland settlements of South Australia, New South Wales, and Queensland, from which transits to the sea become almost unprofitable, ought to be connected by a traversable overland line with the many excellent harbours of the north-west coast; and during my own travels with Mr. Augustus Gregory, in 1855 and 1856, I had recognised the pastoral fertility of many inland tracts of country towards those unsettled places of shipment. If, therefore, an easy overland-route could be found for bringing the surplus stock from the distant inland stations of the eastern portion of our continent to far more remunerative places of occupation and to the very vicinity of safe coast places of export, then an immense impetus would be given to the lucrative settlement of wide tracts of Australia hitherto unnoticed, or even unmapped, and thus gradually palatinal estates might arise by mere small outlays where hitherto blanks only existed on the geographic chart.

Thus the plan was formed to connect the MacDonnell's Ranges with an overland-road to the sources of the River Murchison, or to any other of the great watercourses entering the ocean in more northern latitudes than that stream. I made an appeal through public journals for support of the intended enterprise, but the call at the time was in vain; and as there appeared to be no immediate prospect of geographic parties setting out from any other colony in the direction indicated, the first expedition of Mr. Giles was fitted out at his own private expense and that of the writer, aided by the voluntary service of Mr. Carmichael. The enthusiastic ardour, periodically manifested for territorial explorations, but which seemed for a while to have faded away, was kindled anew when this small private expedition boldly went forth; for ere Mr. Giles had regained the settlements, two other exploring parties, both provided with dromedaries, that of the South Australian Government under the youthful Mr. Gosse, and that of Messrs. Elder and Hughes under the venerable Colonel Warburton, were organised to proceed from middle-stations on the Telegraph-Line also in westerly directions, while soon subsequently West Australia sent eastward her own explorers, Messrs. Forrest, who, like Colonel Warburton, had won renown in former geographic travels. It may be easily understood, that Mr. Giles, having been the first in the field at that period, should have re-entered chivalrously the arena as a champion for geography also; and in as much as space existed, and still exists, for the independent action and movements of several distinct exploring parties, no uncharitable jealousies needed, so we conceived, to mar the successes and achievements of each of these pioneers of civilisation, there being ample room for the work of all.

A second now direct call made by me for aiding so brave and experienced a geographic investigator as Mr. Giles in the renewal and continuation of his journeys met with the generous response of a number of Victorian gentlemen, and the means thus afforded, supplemented by a grant from the Government of South Australia (within whose territory all the first discoveries of Mr. Giles were made), enabled him to start more amply equipped on his second expedition, and throughout this he was most ably assisted by Mr. Tietkens. What can be done by an earnest will, combined with caution and long-trained experience, has been shown by Mr. Giles, and he has by the enterprises already performed stamped his name gloriously as one of the heroes of geography for ever on the pages of Australian history. Modest pretensions and true worth were in these instances nobly joined, and let us trust that under the grace of Providence they will also in his new engagements bring him additional rewards. May we also entertain a hope that those sons of Australia, on whom prosperity has largely smiled, will not forget that they owe tribute to the great land of their birth, and that their gratitude cannot be manifested more gracefully than in fostering enterprises by which at last our Australian Continent will become fully mapped, and at no distant day its far underrated resources may be extensively understood, enjoyed, and utilised.

FERD. VON MUELLER

MELBOURNE, *March, 1875.*

MOUNT OLGA, FROM SIXTY MILES TO THE WEST.

I.

MR. ERNEST GILES'S

GEOGRAPHIC EXPLORATIONS.

1872.

AFTER a rather prolonged sojourn at the Peake (where I met great hospitality, both from Mr. Blood, of the Telegraph Department, and from Messrs. Bagot and Conway, of the cattle station), I arrived at the Charlotte Waters Telegraph Station on the 4th August, 1872, which would be my last outpost of civilisation. I and my companion, Mr. Carmichael, were most kindly welcomed by Mr. Johnston, the officer in charge of this depôt, and by Mr. Chandler, a gentleman belonging to a Telegraph Station further up the line, and in consequence of the kindness we received I extended my stay here to a week. My horses were all the better for a short respite, for the journey up from Port Augusta for several hundreds of miles had been over country nearly destitute of grass and had reduced them to a very low state; from the Strangways, however, the country had been better clothed. The party consisted of myself, Mr. Carmichael and Alexander Robinson, fifteen horses and one little dog. We at length started from the Charlotte on the morning of the 12th August; and, as my intended starting-point to the west was Chambers's Pillar, upon the Finke-River, I proceeded up the line as far as the crossing-place of the last-mentioned water, which is sixty miles by road, though only fifty-three by the line. In the evening of the same day we encamped there. A Government party also arrived under the command of Mr. McMinn, who had completed his professional work for the telegraph line, and was now *en route* for the metropolis. Mr. Harley Bacon, who had been in charge of all the Government stock, also accompanied him, and I had the pleasure of their company at the camp. Close to this crossing-place a considerable tributary joins the Finke near the foot of Mount Humphries; and Mr. McMinn, Mr. Bacon and myself went a few miles up it the following day. Had I found it coming from the west, I had an idea of leaving the line and proceeding up it; but we found its course to be considerably from the south of west. We found a waterhole about twelve miles above the junction, and returned. The country consisted chiefly of open sand-hills well grassed, but the grass was very dry. We retraced our steps, and returned again to the camp. The following day Mr. McMinn and

Mr. Bacon, after making me several kind presents, left for the Charlotte, but I did not remove my camp until the day following. I must here remark, that Mr. McMinn had but recently returned from an exploring trip out to the eastward from the MacDonnell-Ranges, having descended Todd's Creek, and followed it up for some considerable distance until the country became waterless. I obtained from Mr. McMinn before starting, however, the true course and distance from our camp to the Pillar, and at starting again I travelled upon it; to do so, however, I crossed the Finke three or four times, and on the night of the 22nd August I encamped upon the river, having the Pillar in full view.

Friday, 23rd August.—We had an early start this morning, and made straight for the Pillar, which bore nearly north-west by north from camp. The appearance of this feature, I should imagine, to be unique in Australia, and it is not likely that any future explorer will ever discover so singular a monument, wherewith to immortalise either himself or his patron. For a more detailed account of it I must refer the reader to Stuart's own report of it, but cannot pass it by without a brief description. First, then, to approach the Pillar, which can be seen for a long distance from any side except the north, the traveller must pass over a series of red sandhills, mostly covered with scrub, and clothed with that objectionable vegetable production the so called spinifex (*Triodia irritans*), or, as it may be more easily recognised by some, the porcupine-grass. In regions more to eastward this plant usually grows only in mallee; here it obtains alone. The timber near the Pillar is nearly all mulga (a kind of *Acacia*), though a few tall and well-grown *Casuarinas* of a kind, that is new to me (*Casuarina Decaisneana*), are occasionally met with amongst the porcupine-grass. On our route Mr. Carmichael discovered and brought to me a most peculiar lizard, a true native of the soil. Its colour was mostly green. It was armed at all points and joints with stout spines in double rows along its back, sides and legs; they were curved and sharp. On the back of its neck was a thick knotty lump, with one spine from that, and by which I lifted it. Its tail was armed to the point, and of proportional length to its body. It was altogether about seven or eight inches long. I put it into a pouch and intended to preserve it, but it managed to crawl out of the receptacle, and dropped again to its native earth. By this time we were close to the Pillar, and its outline was most imposing. Upon reaching it I found it to be a columnar structure, standing upon a pedestal, which is perhaps eighty feet high, and composed of loose white sandstone, having vast numbers of large blocks loose and lying about in all directions. From the centre of the pedestal rises the pillar, composed also of the same kind of rock, though at its top and for about twenty or thirty feet from its summit the colour of the stone is red. The column itself must be about 150 feet above the pedestal. There it stands (not, indeed, quite alone, as there are other peculiar eminences near), a vast monument of the geological periods, that must have elapsed since the mountain ridge of which it was formerly a part was washed by the action of the ocean waves into mere sandhills at its feet. The stone of which it is formed is so friable, that names can be cut in it to almost any depth with a pocket-knife; so loose, indeed, is it, that one almost feels alarmed, lest it should fall while he is scratching at its base. In a small orifice or chamber of the Pillar I discovered an opossum asleep; it was the first I had seen in this part of the country, and I was not sure, if they inhabited it or not. We turned our backs upon this peculiar monument, and left it in its loneliness and its grandeur—"clothed in white (sandstone), mystic, wonderful." From there we travelled on a bearing of south 87° west, and in seventeen miles came upon some very high sandhills, at whose feet the creek swept; we followed round them to a convenient spot, and one where the horses could water without bogging. I should have mentioned before, that the Finke-River is without exception the most boggy channel I have ever met. As we had travelled several miles in the morning to the Pillar, and the camp was eighteen beyond it, it was late in the afternoon, when we encamped. The country passed over consisted mostly of scrubby sandhills, covered with porcupine-grass, and we passed between some low hills, before we reached the creek; there was a salt channel, where we struck it, and a long hole of brine; there was plenty of good grass on a flat, and we got some tolerably good water near where we fixed our camp; when we had finished our evening-meal the shades of night descended upon us in this our first bivouac in the unknown interior.

Saturday, 24th August.—The mean of meridian altitudes of Vega and Altair last night placed me in latitude 24° 52′ 15″. The night was excessively cold, and the thermometer had fallen by daylight this morning to 18°, our blankets and packs were covered with a thick coating of ice, when we awoke, the tea left in the pannikins overnight had become solid cakes; the water in the creek was running, so it of course did not become frozen. I determined to rest here for a couple of days, as I had many matters to attend to, one of which was to unshoe all my horses, as having now reached a soft country, I would put the shoes away, until I might require them again. From the high sandhills near the camp, as far as I could discern either with the eye or the glass, the creek seemed to be coming from the west. I took some back-bearings upon the Pillar, and decided to travel on a north-west course, when I started again; for although the creek appeared to come from the west, and west was the point I desired to travel upon, yet I could not be too cautious in proceeding, as I felt certain, that eventually it would turn up towards the north or north-west, and I should only get caught in a long bend, and have to go many miles to get out again. The weather to-day was most agreeably warm and pleasant, with a slight breeze from the north-west. We saw the smoke of burning grass, set on fire by the natives to the south beyond a line of ranges, which lay in that direction; the latter had the very red appearance of red sandstone, and they had a series of ancient ocean water-marks along their sides, traceable for miles. I called this Johnstone's Range, after my hospitable host at Charlotte-Waters. There must be some natives about, though we have not seen any since leaving the last-named place.

Sunday, 25th August.—Last night was cold certainly, but not so intensely as the night previously. The thermometer fell to 30°, but there was no appearance of ice or even frost; a very cold puff of wind aroused us from our slumbers about daylight; and this continued from the west all day. The sun threw some rather fiery glances upon us about two and three o'clock; the thermometer rose to 110° in the sun, but in the shade the temperature was agreeable enough. The flies were rather troublesome at times during the day, but after three o'clock they mostly depart. As night approached we could see the brilliant flames of a large grass-fire to the north, about five or six miles distant; this had been started during the afternoon by some prowling sons of the soil.

Monday, 26th August.—Last night was again exceedingly cold, the thermometer at daylight standing at 20°. Some wild animal or other must have been prowling round the camp in the night, for my little dog exhibited great signs of perturbation for several hours. I and Robinson lay awake listening for any sounds, that might give us an idea of our intruders, and we both fancied, that we heard the sounds of human voices; we both got up, and the little dog continued to bark, but we neither saw nor heard anything more. The horses being near, we made an early start this morning, and bore away to the north-west, over a sandhill-country covered with porcupine-grass; it was heavy and distressing travelling for the horses, making them continually jog and jump from the pain caused by the countless prickings, which they received from the sharp thorny spines of the vegetation, they had to pass through. I continued upon this course for eleven miles; and though from the top of a high sandhill, which we reached, we could see nothing whatever of the Finke-River; still I determined to continue another mile or two on this course, hoping I had taken the right direction, and that we should meet the river in a few more miles. We therefore proceeded on, and in two miles we came right upon the top bank of it, the river running under a ridge of high sandhills; its course from this point appeared to incline a good deal to the north. The horses being very heavily packed, and the porcupine-grass having distressed them a good deal, we followed along the bank, and found a convenient spot, where the horses could get water without bogging, and camped after a day's stage of thirteen miles. We passed a few clumps of the fine looking casuarinas, which I have before mentioned; they grow to a height of twenty to twenty-five feet of barrel without a branch, and then spread out into a fine and shady top. They however appear to inhabit the poorest region as far as soil is concerned, for they grow out of pure red sand. A few bushes also grow on these sandhills, and the unfailing—or, I might say, unfeeling porcupine-grass—is almost the only other vegetation. I also noticed a few specimens of a stunted species of *Eucalypt* of the mallee kind. There was a large sheet of water near our camp, and we shot a few ducks, which were settled upon it, when we arrived. The day was agreeably warm and pleasant, with light cool breezes from the north-west. An apparently small tributary from the westward joins the Finke-River to the southward from our camp, and a rather high dark-looking hill forms its most southern embankment. The horizon to the west is bounded by broken ranges of no great elevation; they extend also up to the north-west. The course of the river from here being so much towards the north, our westing is prevented for the present. As we ascend the river the country gradually rises, and we are here about 250 feet above the level of the Charlotte-Waters Station.

Tuesday, 27th August.—The early part of last night was remarkably warm; but towards midnight it became cold enough. The thermometer at daylight had fallen to 28°, but there was neither frost nor dew. Nothing occurred during the night to disturb the stillness of the camp, and we made a start this morning at 8 o'clock. Having discovered last evening, that the river not only trended north, but more north-easterly

I started on a bearing of N. 2° E., and passed over several high sandhills. In two miles we met the river again, almost at right angles. I was unwilling to attempt to cross it, on account of its frightful boggy bed; but, rather than travel several miles roundabout, I decided to try it. We got over it all right certainly; but to see horses sinking bodily in a mass of quaking quicksand is by no means an agreeable sight, and it was only by urging them on with whips, to prevent them delaying, that we accomplished it in safety; the horses, which we were riding, got the worst ground, as it had been so ploughed up by the pack-horses ahead of them. The whole bed of this peculiar creek appears a perfect quicksand; and when my readers are told, that it is nearly a quarter of a mile wide, they will understand, that it is not a feature to be slighted. A stream of slightly brackish water was flowing along its bed, in a much narrower channel than its whole width; and where the water exists, there the bog is most to be apprehended. Sometimes it runs under one bank, sometimes the opposite and sometimes in the middle; a horse may walk upon apparently firm sand towards the stream, when without a second's warning horse and rider may be engulfed in quicksand. In other places, where it is firmer, it will quake for yards all around, and thus give some warning. After crossing, I proceeded upon the same course, having the creek now upon my right hand—over more sandhills and through more triodia for three miles, being five from camp. Here I changed course to N. 17° W., and at one and a-half miles reached a high sandhill. From this point I sighted a continuous range of hills to the north, running east and west; and with the glasses I could see the river bearing up for them. I changed my course for a conspicuous hill in the range, bearing N. 55° W., again having the Finke-River close under our feet and lying right in our course. The alternative of travelling round the long bends was not an agreeable one; and as we had so successfully crossed it once this morning, I decided to try it again. We descended to the bank, and after great trouble in finding a place firm enough and sufficiently large to allow all the horses to stand upon (in the bed), we could not find a place where they could mount the opposite bank, for under it was a long reach of water and a perfect quagmire extending for more than a mile on each side of us. Two of our riding horses were bogged in trying to find a get-away. Finally we had to bridge a place over with boughs and sticks and flood drift, which took us some time. We eventually got them over, one by one, without accident or loss. We then proceeded on our course, and in four miles we touched upon the river again, but had no occasion to recross it as it was not in our road. At three miles further we crossed the dry channel of a small tributary from the ranges to the north of us, near where it formed a junction with the main-stream, which here swept round an elbow towards the west. We found a good watering place in the course of another mile, and encamped, having travelled fifteen miles. The horizon from this camp is bounded from the south-west and west round by north to north-east by ranges, which I am very glad to perceive; those to the south-west and west being the highest and most pointed. It appears, that the creek must come under or through some of those to the north-west. The northern range is distant from our camp about six miles; it appears mainly timbered with mulga. To-day I observed a most beautiful pigeon, quite new to me; it is of a dark-brown colour; mottled under the throat and on the breast. It is considerably smaller than the Sturt-pigeon; it had also a high topknot. It flies in small flocks, and runs along the ground for considerable distances.

Wednesday, 28th August.—The temperature of the atmosphere last night was warmer than usual, and the thermometer did not fall lower than 46°. We made an early start this morning, and proceeded on a north-west course for a pointed and peculiar hill in the ranges to the northward of us, which we reached in five miles, and found the creek flowing at its base. We passed over the usual red sandhills, covered with porcupine-grass, characteristic of the Finke-country; we also passed a shallow sheet of rain-water in a large clay-pan, which is a rather rare feature to meet with in this part of the country. As we approached the hill for which I was steering, it assumed the appearance of a high pinnacle, and the broken fragments of rocks upon its sides and summit showed it too rough and precipitous, to allow a person to climb up it; to the north, the main-line of hills, to which this one belongs, shut out the view in that direction, though I obtained a view of the Finke-River towards the west. I named this peculiar hill after my namesake Mr. Christopher Giles, who is an officer on the telegraph staff at Charlotte-Waters, but was absent on leave at the time of my visit there; and as he is an old Barrier-Range acquaintance of mine I called this hill Christopher's Pinnacle, and the range behind it Chandler's Range, after Mr. Chandler I met at Charlotte-Waters. The creek from here seemed to come from the west, and the line of hills also trended the same way, so I turned my course to N. 80° W.; in two miles we crossed the dry channel of a watercourse, which comes directly from the hills to the N.W. and joins the main-river at this spot; it was broad and shallow, and had very little timber upon it. For some miles we had met with very little porcupine-grass, but here we came into it again, to the manifest disgust of our horses. We were now travelling along the foot of the hills on our right, with the river upon our left hand; in six or seven miles we came to the western termination of Chandler's Range, when we obtained a view to the N. and N.W., where I could see another and much higher range running east and west, parallel to Chandler's Range, but extending to the west as far as I could see. The country hereabouts has nearly all been burnt by the natives; and the horses of course endeavour to pick roads, where the dreaded triodia has been destroyed. We passed a few clumps of casuarinas and a few stunted trees with broad poplar-like leaves. At the end of twelve miles upon this bearing we struck the Finke-River again at right angles, it here running N. and S.; we travelled seventeen miles to-day. I found a fine reach of water here, and the creek has a stony bed so to-night at least, our anxiety as regards the horses bogging is at an end; the stream running over its stony bed produces a most agreeable sound, such as I have not heard for many a long month. Soon after we had unpacked and let go our horses, we were accosted by a cooey from a native on the opposite side of the creek; our little dog became furious, and two natives then made their appearance. We made an attempt at a long conversation, but signally failed, as neither of us understood one word the other was saying; so I shot a hawk for them and they departed. The weather to-day was most agreeably fine, with cool breezes from the N.E.; the sky has been rather overcast, and the flies are most troublesome; it is probable we may have a slight fall of rain before long.

Thursday, 29th August.—The night was cloudy and warm, a few drops fell upon our blankets and made me regret, that I had not had our tents erected; however, no fall took place, so my anxiety with regard to the stores was at an end. When morning broke the atmosphere felt heavy, and there was a sultriness in the air, though the sky was clear; the thermometer fell to 52°, and at sunrise a smoky haze appeared all round the horizon, until it pervaded the whole sky. Whilst we were packing the horses this morning the same two natives made their appearance again, bringing with them a third, who was painted and feathered in the (as they doubtless thought) most alarming manner. I had just mounted my horse and rode towards them, thinking I might get some little intelligence out of the warrior as to the course of the creek, &c., but when they saw the horse approaching them they scampered off, and the bedizened warrior projected himself into the friendly branches of the nearest tree with most astonishing velocity; so I perceived it was useless to attempt to get near them with a horse without running them down, and as I had no desire to do this, I left them. We crossed the creek here on its stony bed, and travelled on a north-west course towards a mountain in the ranges, which traversed the horizon in that direction; it also appeared in the direction from whence the river came. A breeze sprang up from the N.W. also, which caused the dust, raised by the pack-horses travelling upon loose soil, where the grass had all been burnt, to blow directly into our faces. At five miles on this course the creek returned to us in a bend; and at this point we could perceive great volumes of smoke from burning grass rising in all directions; the natives, I suppose, finding it easier to catch game when the ground is bare or covered only with a short vegetation, than when clothed with thick coarse grass or pungent shrubs. At eight miles a tributary creek joins the Finke-River from the north or east of north. It had a broad sandy bed, but was destitute of water, at least where I saw it. At another mile on this course the main-creek swept round to the westward, under the foot of the hills, which we had been gradually approaching. From here I changed my course to west, or indeed a few degrees south of west, and at four miles found, that we had parted with the main-river, and were upon the bank of another tributary, whose junction I had not noticed. The water was running along its bed, but in a very slender stream. It was exceedingly boggy, and we had to pass up a couple of miles before we found a crossing-place, which we at length did. I named this "McMinn's Creek," after Mr. McMinn, whom I have mentioned in the opening of my diary. We turned north after crossing this creek over some stony mulga-hills, and came down upon the Finke-River again in two miles, where we encamped, having travelled seventeen miles. In the evening I ascended a mountain in the ranges to the north of us; it bore N. 33° W.; it was very stony and precipitous, and composed of red sandstone; it was of some elevation, viz., 800 feet above our camp; it had but little other vegetation upon it than huge plots of triodia of the most beautiful and vivid green imaginable, also having the most formidable spines; and whenever one moves they enter one's clothes in all directions, making it truly unpleasant to walk through. From the summit of this hill I could perceive, that the Finke-River turned up towards the hills through a glen, and entered them in a north-westerly direction from the camp. Other mountains appeared in all directions to the north and north-west; indeed this seemed to be a range of mountains of great length and breadth. To the eastward it stretched as far probably as the telegraph-line, and to the west as far as the eye could reach. The sun had gone down before I had finished taking bearings. Our road to-morrow will consequently be up through the glen, from which the river issues. All day a most objectionable hot wind from the west and north-west has been blowing, and clouds of smoke and ashes from the grass-fires, and masses of dust from the loose soil, have made it one of the most disagreeable days ever spent by me. At night, however, a contrast obtained; the wind dropped, and a calm, clear and beautiful night succeeded the hot, smoky and dusty day.

Friday, 30th August.—A meridian altitude of the bright star Vega last night gave me my latitude as 24° 25′ 12″; and though yesterday had been so hot and disagreeable, the night proved cold and chilly; the thermometer fell to 24° by daylight, but there was no frost or even any dew to freeze. Our start this morning was rather late, some of the horses having rambled in the night, the feed at the camp not being very good; indeed the only green herb met by us for some considerable distance is the sow-thistle, which grows to a considerable height, and the horses are extremely fond of it; it is also very fattening. Having at length got under way, we proceeded N.W. for the mouth of the glen mentioned yesterday, through which the river issues. In two miles we found ourselves fairly enclosed by the hills, which close in upon the river on both sides. We had to follow the windings of the channel, which is most serpentine, the mountains occasionally forming steep precipices, overhanging the stream first upon one side, then upon the other. We often had to lead the horses over ledges of rocks, crossing and recrossing the creek perpetually. When we encamped we had only made good eleven miles, though to accomplish this we travelled more than double the distance. We camped close to the junction of a branch-creek, which comes out of the mountain to the westward. There was a good-sized waterhole at its mouth. I have named this "Phillips's Creek," after Mr. Phillips, the telegraph-station master at Port Augusta, who was kind enough to make me a valuable present when passing through that town. This range of mountains is composed of red sandstone, in large or small fragments piled up in the most grotesque shapes. Here and there caves and caverns are to be seen in the sides of the hills. A few trees of the cypress-pine *(Callitris)* were seen upon the summits of the higher mounts. The hills and country generally inside this glen are more fertile than those outside, having real grass instead of the triodia upon their sides. I saw two or three natives just before camping. They kept upon the opposite side of the creek, according to a slight weakness of theirs. Just at the time I saw them I had my eye upon some ducks upon the waterhole; I therefore determined to kill two birds with one stone, that is to say, shoot the ducks and astonish the natives at the same time. I dismounted and got behind a tree (the natives watching me most intently the while) and fired; two ducks only were shot, and the remainder of the birds and the natives (apparently) flew away together. Our travels to-day were agreeable; the day was fine, the breeze cool, and the scenery continually changing; the creek often taking the most sinuous windings imaginable; the bed of it, as might be expected, through such a glen, is rough and stony, and the old fear of horses bogging has departed from us. By bearings upon familiar hills I find our course in a straight line has been nearly N. 23° W.

Saturday, 31st August.—Last night was clear and cold; the stars, those sentinels of the sky, appearing intensely bright, and to the explorer they must ever be objects of admiration and love, as to them he is indebted for his guidance through the untrodden wilderness he is traversing. The thermometer went down to 24° by daylight, but upon the appearance of the sun, the temperature rose rapidly. Several hundreds of pelicans in a large flock made their appearance upon the waterhole near the camp this morning; but no sooner they discovered us than they made off, before a shot could be fired at them; they came from the north-westward, and indeed all the aquatic birds, which I have seen upon the wing, come and go in that direction. Though there are plenty of small fish in the river, yet I do not think they are large enough for mobs of pelicans to exist upon; these birds therefore must have come from some larger waters, I should suppose, up in the tropics; the largest fish I have seen in the Finke was not bigger than a sardine. I am in hopes we shall get through this glen to-day, for however picturesque and wild the scenery, it is very difficult and bad travelling for the horses, and consequently more trouble to get them along. We made but a late start this morning; there was no other road than to continue following the windings of the creek through this mountain-bound glen, in the same manner as yesterday. After travelling some miles, I observed several natives in the glen, ahead of us. Immediately upon their discovering us, they raised a great outcry, made several fires, and raised great volumes of smoke, probably as signals to their friends in the first instance, and to intimidate us in the second, which latter effect did not take place. They then considered it their interest to be off themselves, and they ran away further up the glen; I saw also another lot of some twenty or thirty scudding away over the rocks and hills to our right; they left all their valuables in the camp, which we saw as we passed. One gentleman most vehemently apostrophised us from the summit of a rocky hill, and most probably ordered us away out of his country. We paid, as may be supposed, but little attention to his yells; as his words to us were only wind, we passed on, leaving him and his camp as mere incidents in a day's march. Soon after leaving this native camp we had the gratification to discover a magnificent specimen of the fan-palm,* growing in the channel of the watercourse, with the drift of floods washed against its stem; its dome-shaped frondage contrasting strangely with the paler green foliage of the eucalyptus-trees, which surrounded it. It was a perfectly new botanical feature to me, nor did I expect ever to have met it in this latitude. I had certainly been on the look-out for such an object, as I had noticed portions of palm-leaves and branches in the flood-drifts against the butts of the trees in the glen. This fine specimen was sixty feet high in the barrel. I obtained a quantity of its leaves for my kind and generous friend, the Baron von Mueller, which I brought with me. After passing the palms we continued our march amongst the defiles of this mountain glen, which appears to have no termination, for neither signs of a break nor any-

* A species of *Livistona*, allied to one in the South of Arnhem's Land and now distinguished as the Maria-Palm.—F. v. M.

thing but a continuation of this range could be observed from any of the hills I ascended. It was late in the afternoon, when we left the palm-groves, and in two miles further we encamped; the distance travelled by us since starting was considerably over twenty miles, but I only made good twelve in a straight line from last camp.

Sunday, 1st September.—Last night was bright and cool, but the thermometer did not descend lower than 34°. This being Sunday I made it a day of rest, at least for the horses. I was myself compelled to make an excursion into the hills, to endeavour to discover when and where this apparently interminable glen would cease; for with all its grandeur, picturesqueness and variety of scenery it was such a difficult road for the horses, and so stony, that I was getting heartily tired of it. I climbed some hills to the west of the camp, and passed over cliffs and precipices of red sandstone, and at length reached the summit of a pine-clad mountain considerably higher than any other near it. Its elevation was 1000 feet above the level of the surrounding country. From here I obtained a view to all points of the compass except the west. From here I could descry mountains from the north-east round by north to the north-north-west, at which point a very high and pointed mountain showed its top above the surrounding hills in its neighbourhood; it appeared fifty miles away. To the north and east of north a massive chain, with many dome-shaped summits, were visible: below me towards the camp I could distinguish the channel of the river, where it forced its way under the perpendicular sides of the hills, and at a spot not far above the camp it seemed split into two, or rather I should say it was joined by another creek from the north or little west of north. From the junction the course of the main-stream was more directly from the west. At about ten miles along the course of the tributary I could see an open piece of country, and I thought I discovered with the glasses a sheet of water not far from the foot of the hills, but it was too indistinct to be sure. I was glad to find a break at last in this chain, though it was not on the line I should travel; still I hoped, a few more miles would bring me out on the main-creek also, the course of which I could not distinguish for more than a mile, and that mile was west from the junction. Having expended several hours in my rambles I returned to camp to impart to my companions the result of my observations.

Monday, 2nd September.—Last night was slightly cloudy and warm, the thermometer not falling lower than to 60°. There was a heaviness in the atmosphere, which felt like approaching rain. We started early and proceeded up the glen, still following its mazy windings. In less than two miles we passed the junction of the northern tributary, noticed by me yesterday, and continued on over rocks, under precipices, crossing and recrossing the creek, turning and winding to all points of the compass. One bend perhaps ran west for half a mile, the next turn was perhaps south and so on, so that nearly three miles had to be travelled to make one good. To-day we passed again several clumps of the beautiful palm, growing mostly in the bed of the creek, where they helped considerably to enliven the scenery. I collected also to-day and during the other days, since we have been in this glen, a number of most beautiful flowers, which grow in profusion in this otherwise desolate glen. I was literally surrounded by fair flowers of many a changing hue. Why Nature should scatter such floral gems in such a stony sterile region is difficult to understand, but such a variety of lovely flowers of every colour and perfume I have never met with previously. They alone would have induced me to name this Glen of Flora, but, having found in it also so many of the stately palm-trees, I have called it the Glen of Palms. While we were travelling, a few slight showers fell upon us, giving us warning that heavier falls might be expected. I was most anxious to reach the mouth of the glen if possible by night, so heartily tired was I of such a continuously serpentine track. I therefore kept pushing on. We encountered several natives to-day, but they invariably fled into the fastnesses of their mountain-homes. They raised great volumes of smoke however, and their vociferations only ceased when we got out of earshot. The pattering of the rain-drops became heavier; yet I kept on, hoping at every turn to see an opening, which would form the end, or rather the beginning, of this glen, but night and rain descended upon us, and I was compelled to camp another night in this valley. I found a small sloping sandy and firm piece of ground a little off from the creek, having some bloodwood-trees (peculiar species of *Eucalypt)* growing upon it, and above the reach of any flood-mark. One can never be too careful in selecting a site for a camp on a water-course, for in one night a flood might come and wash everything to destruction. I was very fortunate in finding so favourable a spot, as there was sufficient ground for the horses to feed upon, and some good feed upon it also. By the time we had our tents erected and everything snug the rain fell in earnest; but we were warm and comfortable, and fell asleep in peace and tranquillity, thanking Providence we were much better off than many of our fellow-creatures in the midst of civilisation. The tributary, whose junction we passed this morning, I have named Ellery's Creek, after our well-known and esteemed astronomer, Mr. Ellery, F.R.S.; and the glen through which it meanders into the Finke-River I have named after another scientific and well-known gentleman, Mr. Todd, C.M.G., the pushing Superintendent of the South Australian Telegraph Department. The actual distance travelled to-day in a straight line was eighteen miles, to accomplish which we travelled from morning until night.

Tuesday, 3rd September.—The rain continued at intervals all night, but the showers were slight, and no great quantity of water fell. In the morning the sky was clear towards the south, but to the north dense clouds covered the hills, and the weather was too broken to allow of our travelling to-day. I took another ramble into the hills, to the east of the camp, and upon reaching the first rise, I saw what I was most anxious to see, viz., the end of the glen. It appeared the glen continued for only another mile or two, and the creek then came winding away from the north-west; dense volumes of clouds and mist obscured the view to the north, and it appeared evident that more rain had fallen there, or probably was now falling, than had descended upon us. At mid-day the whole sky became overcast, and more rain fell; up to night it fell in slight showers, but when night had fairly set in it fell heavier.

Wednesday, 4th September.—Rain continued throughout the whole night, but the showers being light no great quantity of water had as yet fallen; an hour after daylight however it came down in more volume, which continued for several hours. At twelve o'clock it held up a bit, and I took the opportunity to plant some seeds of various trees, vegetables, &c., given me specially by the Baron von Mueller. Amongst them were some of the blue-gum tree, cucumbers, melons, many sorts of culinary vegetables, white maize, prairie grass, sorghum, rye grass, &c., &c., also some wattle seeds,* which I soaked, according to instructions, before planting. The rain lasted about thirty-six hours, and altogether nearly three-quarters of an inch fell. Towards evening the sky became clear, and the night was dry though occasionally cloudy.

Thursday, 5th September.—It was with great pleasure this morning that we saw the rain had ceased, and that we should be able to get out of the glen at last. In about two miles after starting we debouched upon the plain which ended at the foot of this chain of mountains, and which I conceive to be the western portion of the two lines of ranges converging into one, named by Stuart the James and Waterhouse-Range. The horses appeared especially pleased to be upon soft ground again. The length of this glen is considerable, as it occupies thirty-one minutes of latitude, which is equal to nearly forty English miles; the main bearing of the glen is nearly N. 25 W.; it is without doubt the longest glen ever traversed by me, and it appears to be the only pass through this range. I have called this mass of hills the Krichauff-Range, after the Honorable Fred. Krichauff of Adelaide, an old university-friend of Baron Mueller. Now that we were fairly out on the open country I found we had a higher and more imposing chain of mountains immediately to the north of us, and that the extent of country between the two lines of ranges was not much more than about twenty miles, if that, though some of the foot-hills of the northern chain came much nearer to us. The northern chain is the western portion of the MacDonnell-Range of Stuart; the river is broader than when in the glen, its bed is stony however still. I fancied I noticed the junction of another from the northward. The country now was pretty level, sandy and thinly timbered; nearly all the bushes had been burnt by grass-fires, set alight by the natives, which appears a perfect custom of theirs. We travelled upon the right bank and cut off the bends, which however were by no means so extensive or so serpentine, as they were on the south side of the glen, or in the glen itself. We met very little porcupine-grass for a great part of the day's stage, as we mostly kept near the creek, but there was abundance of it not far off. The river took us to the foot-hills of the big mountains to the northward of us, and we camped about a mile below a gorge through which it issues. Our course to-day was nearly north-north-west, and we travelled seventeen miles. As we neared the hills we became sensible, that the late rains were beginning to raise the waters of the creek; the sound of it rushing over its stony bed lulled us to repose. We shot a few ducks to-day, six in all; they are very fat and good eating. At about six miles before we camped we crossed the channel of a tributary, joining the Finke-River at right angles from the west. There are also some ranges out in that direction, from whence probably this creek takes its source; the rains have caused a slight flood to take place in it. My next anxiety is to discover, how and where this creek goes next, or if its sources are to be found in this chain of mountains. The day delightfully fine and cool, and the sky clear. The country is rather soft after the rains.

* The wattle seeds given to raise this particular tree for a lasting indication of Mr. Giles's camps.—F. v. M.

Friday, 6th September.—Last night was cold and breezy; the thermometer went down by daylight to 30°. I found my position here to be in latitude 23° 40′, longitude 132° 31′; the variation I previously found was 3° east. I did not move the camp to-day, but took a ramble into the hills to discover the best route to travel upon next. I have reached the foot of a range of mountains, whose eastern portions Stuart called the MacDonnell-Range; and at this part of them they are formed of three separate lines of hills, all running east and west. The most northern of the three are the highest, and indeed are of considerable elevation, rising up to 2000 feet above the level of the surrounding country, and according to my barometrical measurements I found, that at Charlotte-Waters I was 900 feet above the sea; from that point up the Finke-River to the foot of the MacDonnell-Range, I found the country rose over 1000 feet, so the highest points of that range are fully 4000 feet above the sea-level. As I said before, the most northern line is the highest; the other two lines of hills may be called only foot-hills to the higher chain; the most southern and lowest is formed of sandstone, the middle tier is of basalt, and I believe the main chain is of basalt also. After leaving the camp I climbed for several hours over masses of hills, but always found one just a little further on, to shut out the view. At length I reached the summit of a high round mountain, in the middle tier, and a most varied and splendid panorama was spread out before me. To the north was the main-chain, composed for the most part of individual high mounts, with lower ridges between them; there was a valley however between the hills I was on and the main tier, and meandering along through this valley from the west, I could trace by its timber the course of the Finke-River for some few miles. To the east a mass of high and jumbled mounts appeared, and one bluff-faced mount was more conspicuous than the rest. It is probably the Paisley's Bluff of Stuart. Nearer to me and nearly under my feet was the gorge, through which the creek passes, and it appears to be the only pass through the chain. I approached the gorge, and from the top of a cliff I found from the late rains the creek so flooded, that the whole pass was filled with a roaring torrent, so that it would be impossible to get the horses up through it at present, and that the hills which enclosed it were equally impracticable, for it would be next to impossible to get horses over them. The view to the west was gratifying, for the ranges appeared to run on in undiminished height in that direction or a little north of it, and it appeared to me, that the creek must run along this valley for some distance, if it does not take its rise among these hills. From the face of some of the hills, over which I climbed to-day, I saw several springs of pure water running, but if caused by the late rains or permanent, I could not decide. One hill, which I passed over. I found to be composed of what is called pudding-stone, that is to say, a conglomeration of many kinds of stones, mostly rounded and mixed up in one mass, and formed by the smothered bubblings of some ancient and ocean-quenched volcano. The surface of the places, now more particularly mentioned, have been worn smooth by the action of the passage of water, so that it presented the appearance of an enormous tessellated pavement, before which the celebrated one at Bognor in Sussex, which I remember seeing when as a boy on a visit to Goodwood, though more artistically but not more fantastically arranged, would hide its diminished head. I noticed a great quantity of flowers upon the hills in the course of my walk, but as I had collected specimens of the same in the Glen of Palms, and no new ones appeared among them, I did not gather any to-day. I saw also two kangaroos and one rock wallaby. By the time I reached camp it was nearly sundown, and I felt quite disposed for retiring early to rest.

Saturday, 7th September.—I had come to the conclusion in my own mind, that as it was impossible to follow the creek through the gorge in consequence of the flood, and as the hills were impracticable, to fall back upon the tributary I had noticed the day before yesterday, as joining the River Finke from the west, thinking I might in the course of twenty or thirty miles find a gap in the northern range which would enable me to reach the Finke again, should it extend beyond. The night was very cold: the thermometer fell to 28° by daylight. The river had risen still higher in the night, and it was impossible to attempt the gorge. I therefore turned away to west-south-west, in order to strike the tributary. We passed over some rough stony ridges, covered with porcupine-grass. Next we met a thickly bushed sandy country, and struck the creek in about ten miles. After travelling another mile along it I found it trending too much towards the south, as though it came from a different direction to that which I expected; so I camped at a place, where another small dry channel joined it from the north (the new range lying west), and which I expected to be the source of this creek, being too far to reach in one day. There was little or no feed at this camp; not that there was no grass, such as it was—old and dry: the horses will not look at that, having generally had green sow-thistles all along the river, where they grow in the lateral channels. This important western tributary of the River Finke, skirting the south base of the western portion of MacDonnell-Range, has received the name "Rudall's Creek," in honour of an eminent surgeon and promoter of science in the Victorian metropolis, and a friend of Baron Mueller.

Sunday, 8th September.—The night was cold and very dewy, making all the packs and blankets wet and clammy. The thermometer fell to 30°; but instantly upon the sun's appearance it went up enormously. In consequence of the poor feed here the horses rambled, and we made but a late start. I proceeded on a nearly west course direct for the range in that direction. It appeared isolated and of some elevation. I bore for the centre of it. It appeared six or seven miles long, running north and south, and seemed about fifteen miles distant. We passed along the bank of the little creek, which joins this one for a mile or two. We then entered a mulga-timbered country, which ended in a stony ridge covered with mallee and porcupine-grass, which was so dense that only with great difficulty we forced our way into it. It was fortunately only about three miles through, and ended by forming the eastern bank of the creek we had camped on, and which we now met at right angles. The bed here was stony; the slight flood had nearly ceased running, but there were several small pools full where we met it. We watered the horses and proceeded, having travelled ten miles. The day was one of the warmest we had yet felt, and pushing through the scrubs had made both men and horses thirsty. The creek here was trending nearly north and south, and seemed not to come from the range we were steering for; but I thought it only reasonable to suppose, that water would be obtained there so soon after rain. The country after crossing the creek consisted of open high sandhills covered with porcupine-grass; and here I noticed, for the first time in several hundreds of miles travelled to this point, a quantity of the grass-tree *(Xanthorrhœa)* dotting the landscape. They were of all heights, from two to twelve feet. We reached the foot of the range in six miles from the creek. The country round its base is not devoid of a certain kind of wild beauty. A few blood-wood trees, with their brilliant green foliage, enlivened the scene; and, as I noticed a small creek, lined with eucalypts, issuing from an opening in the range, I rode up the glen in search of water; but though I went three or four miles, I was perfectly unsuccessful, as not a drop of the life-sustaining fluid was to be found. On returning to impart this discouraging intelligence to the others, I stumbled upon a small quantity in a depression on a broad sandstone rock, that lay in the bed of the creek. Altogether there was about two quarts. It was too late to return to the creek; and as the horses had so recently obtained plenty of water, it would not hurt them to go without for one night, especially as there was a quantity of herb much like a green vetch or small pea, which they devour eagerly. After supper, which with so small a quantity of water, might be justly termed a frugal meal, I ascended a small eminence near the camp to the north, and with the glasses I could distinguish the creek, last left by us about three or four miles off, now running east and west. I also saw water gleaming in its channel; and more easterly at the spot, where the little creek we were camped upon joined the other, there was also water, but it was equally far away. As the horses were feeding that way, and had already got a mile towards it, I made sure they would follow down the creek and water themselves; but it is very strange, when one wants horses to do a certain thing or feed a certain way, they are almost sure to disappoint one, and so it was in the present case. On returning to camp by a circuitous route, I was fortunate enough to discover in a small rock reservoir an additional supply of water for our own requirements, there being in this last reservoir the enormous quantity of nearly a bucket-full. However it was sufficient for us, and felicity reigned in the camp. This range is composed entirely of red and white sandstone. A few cypress-pines are rooted in the rocky shelving sides of the hills. It is not of such elevation as it appeared from a distance, the highest points being not more that 700 or 800 feet. I collected some specimens of plants, which however are not peculiar to this range alone. I named this Gosse's Range, after Mr. Harry Gosse, who had been out from the Alice Springs Telegraph-Station; but where his travels took him to I did not hear. It seemed, that the late rains had not visited this isolated mass. It is barren, and covered with triodia from turret to basement, wherever the stones are not too thickly crowded to prevent its growth.

Monday, 9th September.—The night, like the preceding one, was cold and dewy, and the thermometer fell to 30°. The horses, it seems, wandered quite in the wrong direction during the night, and it was 11 o'clock before we got away from camp. We went north for the sheet of water, seen by me from the top of the little hill near the camp in the creek about three miles off to the north. Upon arrival we watered the horses and proceeded up the creek, as its course here appeared to be nearly from the west. The country along its banks was level, open, sandy, and covered with the widely pervading porcupine-grass. We passed several grass-trees again to-day. At various distances several small tributaries joined the creek from the ranges to the north of us, but I saw no water in any of them. We found water in the main creek all along its channel; but I do not think we should have been so fortunate had it not been for the late rains. Our course to-day was nearly west by north, and the distance thirteen miles in a straight line. The weather was rather warm to-day, and I did not feel comfortable until I abandoned my coat. The lines of foot-hills of the northern range are still running parallel to our course, and they hide from our view the country in that direction.

Tuesday, 10th September.—Last night was considerably warmer than any we have experienced as yet, except perhaps the night of Sunday, the 1st. The thermometer fell no lower than to 50°. We made an early start this morning, and continued along the creek, which we had been following for the last two days. At four miles I found it turned up to the north-west, then to the north, and it became confined on both sides by stony mallee-hills, forming a small glen. I continued up the glen for two or three miles, thinking it might open out; but such was not the case, and I was compelled to leave it, and ascend the mallee-hills upon its western bank. The hills from whence the creek issues were not many miles away to the north, and the country between us and them was all stony and scrubby. I now travelled about west-north-west over mallee-hills of the most wretched and frightfully barren description. The scrub was very thick, and we had very difficult work to keep the horses in a body at all, as each of them endeavoured to find an easy path for himself. This continued for several miles until we reached one ridge higher than the rest, and which I found to be the highest our horses had passed over since leaving Charlotte-Waters. From here with the glasses I fancied I saw the timber of a creek in a valley to the north-west; at all events I could see the country was not stony mallee; so I made off in that direction, and soon came upon the channel of a small dry watercourse, lined with eucalypts, which was the first I had met, whose waters flow towards the west; not that there was any water when I saw it, but the flood-drifts against the timber on its banks indicated the direction of its fall. I followed the creek down, hoping to meet with water; but it ran into another and rather larger one, in the bed of which, after following it for a mile or two, I found a small pool. The water had evidently lain there many months, as it was half slime and drying up fast; it was evident that the late rains had not fallen here. We were travelling for a long time to-day, and made a stage of nineteen miles; our course was a little west of north-west, though we made traverses upon nearly all the northern points of the compass during the day. The weather was quite warm enough, and when we encamped we felt the benefit of what shade the creek timber could afford. There was good feed here for the horses—some of the small vetch or pea-like plant they are so fond of. The grass is useless, being old, long and withered. To-day, in the glen formed by the creek we encamped upon yesterday, I saw a single quandong-tree in full bearing, but the fruit not yet ripe. I also saw a pretty drooping kind of acacia-tree, whose leaves hang in small bunches together, giving it an elegant and pendulous appearance. It grows to some height; two or three I saw must have been forty or fifty feet, and over a foot through. The flies to-day were exceedingly troublesome. I have not seen much game lately, with the exception of a few emus. I noticed one so-called native orange tree *(Capparis)*, of a very poor and stunted habit.

Wednesday, 11th September.—The night was again remarkably warm, and the thermometer fell only to 52°. My latitude last night I found to be in 23° 29′ S., which places us almost under the tropic line. The horses having remained within sight of the camp all night, I made an early start this morning, intending to follow this new creek, which received the name of my companion, Mr. Carmichael, and I hoped it would take me some considerable distance on my road to the west. It soon appeared its course was considerably to the south of west. However I followed it for several miles, when it turned up again to the north-west. The main line of mountains we had still upon our right, or north of us; and to the south of us another line of lower hills trended up in a north-westerly direction towards them, and towards the west or west-north-west there seems a kind of gap between the two lines of ranges about twenty-five miles off; I hope this creek will carry us that far. The country along the banks of the creek was open and sandy, with plenty of old grass and not much triodia, but to the south the latter and mallee seemed to approach very near. We saw several small ponds in the creek as we passed along, but none of any size. Where the creek turned up to the north-west, the mallee upon the south bank gradually encroached, and at length covered its southern bank. I had only followed it seven or eight miles, when I found it showed signs of falling off, and split into several channels, and eventually ran out on an open swamp or plain. Another small creek from the north joined before this finally exhausted itself, and they both ended on the plain, and there was no water to be got in either channel. The little plain looked green and most agreeable. I found some rain-water in clay-pans upon it; and as the feed was so excellent, I encamped upon it. There were numerous kangaroos and emus on the plain when I arrived, but they did not remain in our company long. I noticed in the mallee on the south bank of the creek to-day a tree I had not before seen; its bark was dark like mulga, but its leaves were like the curragong. I brought some away with me. My course to-day was nearly west, though I travelled south-west, north-west and north to reach the spot where we encamped, distance fourteen miles.

Thursday, 12th September.—I remained at this camp to-day for the benefit of the horses, as the grass was excellent; indeed this is the most agreeable and fertile little spot I have met for some time. It consists of a small plain, bounded upon the north by the most peculiar looking mountains. It is also fringed with scrub nearly all round. The appearance of the northern mountains is most singular and grotesque, and very difficult to describe. There appear to be still three distinct lines, one of which ends in a bluff, to the east-north-east of the camp; another line ends likewise in a bluff to the north-north-east, named Haast's Bluff, after Dr. Haast, F.R.S., geologist, Canterbury, New Zealand, an alpine explorer; the third continues along the northern horizon. One higher point than the rest in that line bears N. 26° W. from camp. The middle line of hills is the most strange looking. It recedes in the distance eastwards in almost regular steps or notches, each step or notch being a bluff itself, and all overlooking a valley. The bluffs all have a circular curve, are all coloured red, and in perspective appear like a gigantic flat stairway, only that they have an oblique tendency to the southward, caused, I presume, by the wash of the ocean currents that at perhaps no greatly distant geological period must have swept over them. My eyes however were mostly bent upon the high peak in the northern line, and Mr. Carmichael and I decided to walk over to ascend it. We started at two o'clock, and though it was apparently not more than seven or eight miles off, it was nearly midnight before we again reached the camp. As the reader is aware, I left the River Finke issuing through an impracticable (at the time) gorge in these same ranges, now some seventy-five miles behind me, and during that distance not a break had occurred in the lines of mountains, whereby I could either get over or through them, and thus meet that stream again; indeed, at this distance it was doubtful, if it were worth while to endeavour to do so, as one can never tell, what change may take place in the course of a stream in that distance. When last time seen it was trending along a valley under the foot of the highest tier of hills, and coming from the west; but whether its sources are in those hills or it extends still west and somewhat north of my present position, that is to say on the other side of the chain of hills to the north of me, is a question, which I hope to solve by ascending the high point about seven or eight miles off. I am the more anxious to reach the River Finke (if it is still in this region), as on the route along which lately I have been travelling, water has been by no means plentiful, and I believe that a better country altogether exists upon the other side of these mountains. At starting for the mountain Mr. Carmichael and I passed over at first the plain upon whose southern edge we were encamped. It was beautifully grassed, having good soil upon it, and it would make a fine (bush) racecourse, or be an excellent spot for hunting kangaroos and emus, of which we saw a great number. In about three miles the plain ended in a thick, indeed very dense, mulga scrub, which continued to the foot of the hills. The grass was long, dry and tangled, with dead and burnt sticks and timber, and it was exceedingly troublesome to walk through it. Upon reaching the hills I found the natives had recently burnt all the vegetation from their sides, leaving the stones, of which it is composed, perfectly bare. It was a long distance to the top of this first ridge, but the incline was easy, and I was in great hopes if it continued like this to be able to get the horses up over the mountains at this spot; but upon arriving at the top of this hill I was soon undeceived upon that score; for the high mountain, for which we were steering, we found completely separated from us by a yawning chasm, which lay under an almost sheer precipice at our feet. The higher mountain beyond was girt around by a solid wall of basalt of fifty or sixty feet in height, from the top of which the summit rose. It was quite unapproachable, except in one place round to the northward of it. The solid rock, of which it had formerly been composed, had by some mighty effort of Nature been split into innumerable fissures and fragments, both perpendicularly and horizontally, and was almost mathematically divided into pieces or squares, simply placed one upon another like mason work without mortar; the lower strata of the hill being large, the upper ones tapering to pieces not much larger than a brick. The whole appearance of this singular mountain was grand and awful, and I could not but reflect upon the time when these colossal ridges were all at once rocking in the convulsive tremblings of a mighty volcanic shock, which shivered them into the fragments I now beheld. I said the hill we had ascended abruptly ended in a precipice. By going further round to our left we found a spot, which was practicable but difficult enough to descend. At the bottom of some of the ravines below I could see several small pools of water gleaming in little stony gullies. The afternoon had been warm, not to say hot, and our walk and climbing exertions had made us thirsty, and the sight of water made us all the more so. It was now nearly sundown, and it would be useless to attempt to ascend the high mountain, as by the time we could reach its summit the sun would be far below the horizon, and we should obtain no view after all. It was however evident, that no gap or pass existed, by which I could get my horses up, even if the country beyond were ever so promising. A few cypress-pines dotted the summits of the hills, and they also grew on the sides of some of the ravines. We had, at least I had, considerable difficulty in descending the almost perpendicular face of the hill to the water we saw below. When I reached it Mr. Carmichael had had time to lave his feet and legs in a fine little rock-hole full of pure water, filled, I suppose, by the late rains. The water, indeed, had not yet ceased to run, for it was trickling from hole to hole. Upon Mr. C. inquiring what delayed me so long, I replied, "Ah, it is all very easy for you, you have two circumstances in your favour—you are young, and therefore able to climb; and besides you are in the tropic." To which he very naturally replies, "If *I* am in the tropic, *you* must be also." I benignly answer, "No, *you* are in the *tropic clime of youth.*" No view of any kind except along the mountains for a mile or two east and west could be obtained. I was greatly disappointed at having such a toilsome walk for so little purpose. We returned by a more circuitous route down a small water channel; but soon had to take to the scrub, and eventually reached the camp at nearly 12, thoroughly tired out with our walk. I have since named the high mountain, which I did not reach, Mount Musgrave, after His Excellency the new Governor of South Australia. It is of considerable elevation, being 1600 feet above the level of the surrounding country, and over 3000 above the sea.

Friday, 13th September.—Having many small matters to attend to in camp, I remained here another day. The thermometer fell only to 50° last night. This morning there was a very cold breeze, which we all felt keenly, probably because the last few days have been so warm. The sky all day has been overcast, and there is a feeling like approaching rain in the atmosphere. There were several emus and kangaroos near this camp, and there were also some crows, hawks, quail and bronze-winged pigeons. Mr. Carmichael went out to-day to shoot a kangaroo, but it appeared he was unsuccessful.

Saturday, 14th September.—The thermometer at daylight was rather high, standing at 55°, warning us of the approach of a warm day; the flies were soon very busy at our eyes, and soon after sunrise it became hot. We got away early from camp, and proceeded across the little plain in the direction of another high bluff-faced mountain, which loomed over the surrounding country, to the west-north-west. In three miles we had traversed the plain and entered a mulga scrub, which had been recently burnt by the natives: the further we proceeded through this the worse it got. At seven miles we came upon mallee, stones and triodia; the mallee we found so dense, that not a third of the horses could be seen together, and it was with the greatest difficulty we managed to get through it to the foot of a small pine-clad hill, lying under the foot of the high bluff before mentioned. There was a small creek, lined with eucalypts, which comes down from the high mountains, and runs under the little pine-clad hill. We only travelled twelve miles, but that distance through such a scrub took many hours. We passed three other small creeks before we reached the pine hill; they all run south-west, and form a larger one, which I could see from the top of the pine-hill, about two or three miles off. Mr. Carmichael went up this little creek and found a small rocky waterhole, with barely sufficient water for our use. Here I encamped, as it was very acceptable, the day had been disagreeably warm, and working in the scrub had made both men and horses thirsty. I left Mr. Carmichael and Robinson to unpack the horses, and rode over to the creek I had seen to the south. I found two small puddles in its bed, but not sufficient for the requirements of our horses; there was however plenty to be got by digging, as the bed of the creek was damp and wet, and by scratching with my hands I soon got some. The camp was fixed in a wretched place, in the midst of dense mallee and thick plots of trinodia, which we had to cut away before we could sit down. We could not see a yard ahead of us in any direction except that of the sky, where the big bluff was observable. I was sorry to find, that the horses had been let go without hobbles, and as they had been in such fine quarters for three nights at the last camp, it was more than probable they would make back through the scrub to it, and on the morning of

Sunday, 15th September, not a horse was to be found. The night was sultry, and the thermometer did not descend lower than 60°. Robinson and I went in search of them, we found they had split up into several mobs. I found three in one lot, and at night Robinson returned with only six more; the remainder had been missed in the dense scrub. The day was exceedingly warm, the thermometer stood at 95° in the shade, and there was also a warm wind blowing. Robinson and I started on foot, and as he had to go back to the last camp before he found his mob, and then drive them through that wretched scrub by himself, he had a fine day's work of it. I got my three much closer at hand, and after returning to the camp in the afternoon I attempted the high bluff immediately overlooking the camp. I had a great deal of cliff-climbing, and reached the summit of one mountain of considerable elevation, it being 1300 feet, and then found that a vast chasm or ravine separated me from the main mountain-chain; it being near sundown when I reached the summit of the one I was on, and as it would have taken me two or three hours to get up to the summit of the highest, it would be too dark to get any view; I did, therefore, not attempt it. It was of great height, as it towered above the hill I was upon, and was 500 or 600 feet higher than it. By the time I reached the camp Robinson had but just returned. I collected a few botanical specimens at this camp for my generous patron Baron von Mueller. The mountain I ascended and the higher bluff were composed of basalt; very little timber existed upon them, but they were clothed with porcupine-grass to their summits; several beautiful flowers grew upon them, which I collected.

Monday, 16th September.—The night was again sultry and cloudy, the thermometer not falling below 60°; it rose rapidly with the sun, and it was evident, that we were to have another hot day. Robinson and I went again on horseback after the missing horses. We had to return through the scrub to our old camp upon the plain, and then we found them all with the exception of one, which was still absent; by the time we returned with them to camp it was evening again. The weather having been very hot, the thermometer in the shade at the camp had stood at 96°. Not hobbling my horses in general, we had some difficulty in raising a pair for each horse, and not being able to do so, I left one in the mob without, and the consequence was, that the ungrateful creature crept away by himself.

Tuesday, 17th September.—The night was again exceedingly sultry; the thermometer stood at 62° at daylight. As this camp was the most wretched hole it was possible for any man to have got into, being in the midst of dense mallee and triodia and stones, I determined to escape from it before attempting to look for the two still missing horses; and as there was the other creek two or three miles away, which the one we were on emptied into, I packed up and went to it, three miles S.W. In fact the water at the camp was completely done, as there was only a small quantity when we first camped at it, and on that account alone I should have been compelled to leave it. I camped upon the new creek at the little holes I had seen before, but the surface water had almost disappeared. We therefore had to dig out the sand, and thus we obtained a sufficiency by a great deal of labour. In the afternoon Robinson and I again went to look after the two missing horses. We followed up this creek and found some water in it three miles above the camp; presently we found the tracks of our horse, and found he had been about there for a day or two, as the tracks were old. We made a sweep out round some hills and found the tracks again, much fresher. Eventually we found the little horse, feeding by himself, six or seven miles up the creek, above the camp. It was too late then to go further to look for the other horse, so we returned to camp. The day was again very hot, the thermometer at the camp having gone up to 96°. The sky was very much overcast, and rain appeared imminent.

Wednesday, 18th September.—In the course of the night some drops of rain fell; it was cloudy and close, and the thermometer at daylight stood at 61°. I sent Robinson away back to our old camp upon the little plain, feeling pretty sure that as the missing horse had returned there once he would do so again; and as he had had plenty of time to get through the scrub I thought it probable Robinson would have only to ride there to get him. There was a hot wind blowing to-day from the N.W. The sand was flying about in all directions, and it was a most uncomfortable and disagreeable day; the thermometer indicated 96° in the shade, and I was only too glad when night approached, as the wind seemed inclined to lull. Robertson returned in the evening, having been successful in finding the horse at the old feeding ground, and I took special care to find a pair of hobbles for him for this night at all events. The atmosphere was heavy and overloaded, and it seemed probable, that a fall of rain would occur, as the heat experienced by us for the past few days was quite unseasonable; the flies also were most insufferably pertinacious, not that they were so numerous, but I never met even amongst their congeners upon the Darling anything like such persistence, and I therefore consider the tropic fly of Australia the most abominable insect of its kind. At this camp we were about five miles from the high bluff, and which was now to the east of north from us. From the top of the mount, ascended by me on Sunday, I found the line (now called the Liebig-Mountains, after Baron Justus von Liebig, from whose great chemical researches Australia has also already so much benefited) of mountains still ran on towards the west; they would not intercept the horizon at that point exactly, but they would so at west by north. The farthest ranges I could see from the hill appeared forty or forty-five miles distant, and they took a sweep or curve round in a northerly direction; the most distant one I saw appeared still high. I could not discover the conformation of the whole line on account of the curve which they took, and which some nearer high points also intercepted. However, as they extend forty miles, I shall still follow them in hopes of meeting some creek or other ranges, which will carry me on towards the west. It is a most remarkable fact, that such high mountains, whose feet I have been travelling along, should send out no creek whose course extends longer than ten or twelve miles. The creek upon which I am now camped I could trace by its timber for a few miles; its course was at first nearly west, then it appeared to turn more to the south. The country in its immediate neighbourhood is open and timbered with fine casuarina-trees; the grass is long and dry, and the triodia approaches within a quarter of a mile of the banks. I have now reached the farthest or most western point of the line of hills, which we have had for some distance upon our left or to the south of us. I named this line of hills Gardiner's Range, as Mr. Carmichael desired me so to do, after a friend of his. Though there is still one small isolated hill, the furthest outpost of the line, some three miles away to the south-west, the creek may probably take a bend down under it. This creek is rather well timbered, the eucalypts look fresh and young, and there is some green herbage in places, though the surface water has all disappeared.

Thursday, 19th September.—The night was again close and warm. Rain seems imminent, but still it hangs off. The thermometer did not descend below 60°, and soon after sunrise a disagreeable feeling of oppression and mugginess was in the atmosphere. All the horses were found together this morning; but in consequence of there being so little water at the camp, I had to send them to the little waterhole I had found three miles up the creek; therefore it was late by the time they were packed, after going there and back. I was not sorry to be moving again; for our stay at these last two camps had been compulsory, and the anxiety, trouble and work we had had here left no very agreeable reminiscences of the locality in our mind. I followed along the creek all day, cutting off the bends: but though we travelled some thirteen or fourteen miles, I saw no signs of any water. As it was getting late, I decided to try if we could obtain some by digging. We set to work, and very soon got some wet sand, and we got water in about four feet; but the sand being so loose, we had to remove an enormous quantity to enable a horse to approach. Some horses of course would not go near it, and had to be watered with a canvas-bucket. The supply of water seemed good, but it only ran in from the sides. Depth was no object, as it did not appear to come from the bottom, which gave us all the more work, as after every horse had drank, we had to dig afresh. The country was open to-day, and timbered with the fine casuarina; triodia reigned supreme, however, within not half a mile of the creek. The old grass had been burnt near this place, and good young green shoots appeared in its place, which were very good for the horses. The day was close and sultry, and the rain seemed to have departed entirely, though the sky was still clouded, especially towards the west.

Friday, 20th September.—The early part of the night was hot and cloudy, but it eventually became cooler; and towards morning a few drops fell. The thermometer stood at 55° about daylight, and we saw flashes of lightning in the west, and heard the dull roll of distant thunder. While we were at breakfast a thunderstorm came up; but it passed immediately close to us on the south bank of the creek, only just sprinkling us. I thought it had entirely departed, so I decided to pack up and be off, hoping to find a better watered region in the neighbourhood of the line of mountains, which I was now intending to visit. There was an extraordinary mount a little to the west of north from us; it looked something like a church at about twenty miles off. I called it Mount Peculiar. We had great trouble again in watering the horses; but eventually they were all satisfied, and we left the camp behind us at last, leaving the creek also on our left to run itself out into some lonely flat or dismal swamp, known only to the wretched creatures of this desolate region. My course to-day was about N. 67° W. over a sandhill country, rather open, covered with triodia and timbered with casuarina. We had scarcly gone two miles, when the thunderstorm returned, and the rain fell rather heavily. The country however was so sandy and porous, that none remained upon the surface. I had no alternative but to travel on, hoping to find some spot, where it might lodge. The rain continued to fall heavier and heavier; still we could not stop, for not a drop of water was to be had, the ground sucking it up as fast as it fell. We continued our voyage, and travelled on this course twenty-five miles, it raining hard nearly all the time. At this distance we came in sight of the mountains I was steering for; but they were too distant yet, to be reached before night; so I turned up a little northward to the foot of a low bare white granite hill, hoping to find a creek or some ledges in the rocks, where we might get some water; but none was to be found; thus we had to camp without, although we had been travelling in the rain the whole day and accomplished thirty miles. There was good green feed here, where the old grass had been burnt; and as it was still raining, the horses could not be very much in want of water. We fixed up our tent, and retired for the night, the wind blowing furiously, which of course might be expected, as it was the eve of the vernal equinox, and this, I suppose, is our share of the equinoctial gales. It rained once or twice only in the night, but not very heavily. The last few miles traversed by us were through a thick mulga-scrub, which ran up to the foot of the hill near which we camped.

Saturday, 21st September.—This morning was dull and cloudy, and the thermometer was still rather high, standing at 55°. As there was no water here, I was compelled to leave this camp, rain or no rain. Indeed, unless it descended in perfect sheets, the country would not hold it, being all pure red sand. The hill near us had no rocky ledges to catch water, so I made off for the higher mountains, for which we were steering yesterday; but now I determined to make for their nearest and or most eastern point, which was not more than four miles off. I did not wait for the horses, but walked away immediately after breakfast with the shovel, thinking I might find some place, where water might be got by digging, having pointed out a hill to Mr. Carmichael, to bring the horses to when packed. From the hill near the camp, with the glasses, I noticed a few eucalypts apparently in a small channel issuing from the ranges I was approaching. I therefore went to it, and followed it up to the rocky ledges, from which it obtains its source, and down which at times, no doubt, leaping torrents roar. Very little of the rain had fallen here, but I found most fortunately one small rock reservoir with just sufficient water for all the horses. There was none either above or below it in any other basin, and there were plenty of better-looking holes, but all dry. The water in this one must have stood for some time, yesterday's rain not having affected it in the least. Where I found the water was in the most difficult place for horses to reach; indeed it was almost impracticable. After finding this opportune supply, I walked or rather climbed to the top of the mount, on whose sides I had found the water. On the summit was a native fig-tree in full bearing. The fruit was ripe and delicious; it is about the size of an ordinary marble—yellow when unripe, gadually becoming red, then black: it is very like the English medlar, only rather sweeter, and full of small seeds. I was disturbed from my repast; but seeing the horses, several hundred feet below me, going away in the wrong direction, I had to descend before I had full time to look around me; but the casual glance I obtained from the top of the hill gave me the most gloomy and desolate view imaginable, and almost enough to daunt the explorer from penetrating into such a dreadful region. To the eastward I discovered, that I had long outrun the old main-chain of mountains, which had turned up to the north, or rather north-north-westward; between me and it a mass of jumbled and broken mounts appeared, each separate hill almost surrounded with scrubs, which came up to the foot of the hill I was upon. To the north the view was very similar, some isolated and some joined ranges appeared, but all having dense scrubs between them. To the west the picture was the same, ranges loomed above the intervening scrubs. Those further to the west appeared a considerable distance away—probably fifty miles—and they also appeared to be of the greatest elevation. The country between was also filled in with ranges; the whole horizon looked dark and gloomy. I had to abandon my scrutiny of the country however, and descend to the horses. There appeared no creeks of any kind, the most extensive not running more than a mile from the hills it might issue from. Though no doubt there is an ample rainfall in this region, the water sinks into the earth as fast as it falls from the heavens. Watering the horses proved a difficult and tedious task, as many of them would not approach the little basin, in which the water lay; it therefore had to be carried to them in buckets. By the time they were all watered, and we had descended from the rocky gully, it was time to camp, the day having passed with miraculous celerity. The horses fortunately did not drink all the water; indeed there was nearly sufficient to give them another drink. The feed was very good here, as there was a little flat, lately burnt, and it was covered with yellow flowers, which the horses like.

Sunday, 22nd September.—The night set in very cool, clear and calm.

I was enabled to take a meridian altitude of Vega, which placed me in latitude 23° 15′ 12″ S. We were distant now from the last water in the creek nearly thirty-five miles. The mercury in the thermometer fell as low as 28° by daylight this morning, rendering the temperature much colder, than we had experienced it for some time; there was also a slight frost. I determined to rest the horses here to-day, and let them drink up what water was left in the rocks, whilst I went away to inspect some other gorges or gullies in the hills to the west of us, to see if any more water was to be found. The day was most agreeably cool and fine, and I set out in great expectation of finding a fresh supply. I climbed to the summit of one of the highest hills in the range, and found it was 800 feet from the base. The view obtained from its summit was nearly similar to that mentioned yesterday, only I could see rather more of the horizon towards the west. It appeared that the ranges I was now on ran on for about twelve or fourteen miles; then a lot of low jumbled ridges were visible above the scrubs, and the farthest range to the west or rather west-north-west appeared rather more inviting than before, but the intervening country was a mass of dense scrub; a few low ridges also appeared here and there to the south, and an odd flat-topped tent-shaped hill showed about N. 30° W., some twenty-five or thirty miles away. No sign of any creek appeared to issue from any of those hills, except indeed little gullies, having an odd eucalypt here and there in them, and their course seldom reached a greater length than one mile, before they were swallowed up in the scrubs; most of their little channels were not more than six feet wide, and the trumpery little streams, which descend from them in their most flooded state, would be of little service to anybody. I had travelled over hills and through gullies, and up mountains all the morning, yet I had not met with a single drop of water, and was returning disappointed to camp, when, on pushing up another small, scrubby, dreadfully rocky little gully, which I had missed in going out, I was at last fortunate enough to find a few small rocky holes, full of the purest fluid; but what pleased me most was the rather strange fact, that the water was running from the one into the other, but the stream was so weak, that I should more properly call it trickling. Above and below where I found the water, the gully and the rocks were as dry as the desert around. Had the water not been running, and thus the supply kept up, half my horses would have emptied all the holes at a draught. The approaches to the water were even more rough and impracticable than those near the camp. I was however most delighted to have found it, otherwise I must have retreated to the last creek. Having however found this supply, I determined not to touch upon it so long as I had a drink at the camp for the horses; and in the hills farther west, to which I intended to travel, I hoped to get more. So I kept this small supply as a reserve to fall back upon, should I be unfortunate in obtaining more further on. Returning to camp, I got all the horses together and took them up to the last watering-place, and by the time they had all drank—some from canvas buckets, and some from the hole—there was not a drop left in the whole gully. It was therefore imperative upon me to be off, as this supply was gone.

Monday, 23rd September.—The night was again cool and clear, and the thermometer stood at 30°. A heavy dew fell, but no frost appeared. The horses wandered a bit in the night, and it was late in the day when we got away from the camp. I proceeded along the foot of the ranges, which here ran nearly W.N.W. Close to their foot the country is full of stones, but open and thickly covered with enormous branches of triodia, which by this time the horses dread like a pestilence; I have encountered this scourge for more than 200 miles. All round the coronets of most of the horses, in consequence of being so continually punctured with the spines of this vegetable, it has caused a swelling or tough enlargement of the flesh and skin, giving them the appearance of having ring-bones, many of them having the flesh quite raw and bleeding; they are also very tender-footed, from traversing so much stony ground as we have had to pass over. Bordering upon the open triodia ground, to the south of us, is the scrub, composed chiefly of mulga, though there are various other shrubs and bushes growing amongst it; it is so thick, that we cannot see one-third of the horses at once; they of course continually endeavour to make for it, to avoid the stones and triodia, for generally speaking, the pungent grass and mulga appear antagonistic plants; the ground however is usually soft in the scrubs, and on that account the horses also seek it. I have occasionally allowed them to travel in the scrub out of kindness to them, until some dire mishap forces us out again; for the scrubs are mostly so dense, that the horses are compelled to crush through it, tearing the coverings off their loads, and occasionally—indeed frequently—forcing sticks in between their backs or sides and their saddles; then we hear a frantic crashing through the scrub, and the sound of the pounding of horse-hoofs being the first notice we receive, that some frightful calamity has occurred; then, as soon as we can get through it ourselves, and round up the horses, we find one missing, whose tracks have to be found; then portions of the load are picked up here and there, and perhaps in the course of an hour the horse is again found and repacked, and we push on again. Sometimes it is varied by there being two horses that have disencumbered themselves instead of one; when those accidents occur we push out again immediately for the open though stony ground, for there at least we can see what is going on. These scrubs are really dreadful, and one's skin and clothes get torn and ripped in all directions. After these mishaps, one of which occurred upon this day's travels, I continued on and travelled ten miles, to the foot of a hill I had been steering for. There were numerous gullies or miniature-creeks coming from the hills. A few cypress-pines were scattered here and there upon them, the hills themselves having a glistening, sheening, laminated appearance, caused by the vast quantities of mica, which abound in them, their sides being furrowed and corrugated, and the upper portions of them almost bare rock. I made a search for water amongst the gullies, and expended some time in doing so, but was quite unsuccessful. Another range of the same kind lay three or four miles further on, apparently higher, and therefore, I supposed, more likely to have water near it. Accordingly we went to it, and halted the horses, while Mr. Carmichael and I ascended to the summit of the highest, which I found was 900 feet above the level of the surrounding country. We had already searched all round its base for water, and found none. The view from the summit was such as I have described before, that is to say, an ocean of scrubs with ranges appearing above in most directions. It gave however no hopes of any water. The horses had already been twenty-four hours without any.

I saw that it was useless to push further with all the pack-horses through the vile scrubs, which intervened between this and the ranges further to the west-north-west. One point of the ranges in that direction was high and pointed, and I decided before abandoning this line of march to visit it; but I must first find a place to leave the main body of the horses under the care of my companions. The ranges in question seemed twenty-five or thirty miles away. No watercourse was visible in any direction, except indeed the little runnels, that come down from all the hills here and there, and which, as I have before remarked, scarcely run a mile, before they become lost in the scrubs. To the south the country seemed to have been more recently visited by the natives than any other part, as burnt patches could be distinguished as far as the vision could be carried; and in the far distance, over the top of some low ridges (probably forty miles away), we saw the ascending smoke of grass fires still attended to by the natives. I had no alternative but to retreat upon the small supply of water I had left behind me as a reserve, so I turned the horses' heads again in that direction. This time we travelled upon the northern side of the hills, which previously we traversed on the south side, to make sure of not missing any creek, which might flow from them. We camped in a grove of casuarinas, where the old grass had been burnt, and some young stuff was springing, but of course there was no water. We have been so short of that commodity for some days, that the merest attempt at a wash has been out of the question.

Tuesday, 24th September.—A fine dewy night made me hope, that the horses would not feel the want of water so much, as to induce them to ramble; but somehow this appeared foredoomed to be an unlucky day in my calendar, for it really appeared as if everything must go wrong by a natural law. In the first place, while making a hobble-peg, when Robinson was away after the horses, the little piece of wood slipped out of my hand, and the sharp blade of my knife went through the top and nail of my third finger and stuck in the head of my thumb. They bled most profusely, and it took me until the horses came to sew my mutilated digits up. It was late, when we retreated from camp. As there was a hill with a prepossessing-looking gorge, I left Mr. Carmichael and Robinson to bring the horses on, and went to see if I could possibly find water in it. I went some distance up through the stones and ridges; however there was none. I then made down to where the horses should have passed along, and found some of them standing in a small bit of open ground surrounded with dense scrubs, with nobody near them. I cooeed and waited, and at last Mr. Carmichael came up and told me that when he and Robinson with the horses debouched upon this little opening they found that two of the horses were missing, and that Robinson had gone to pick up their tracks. The horse carrying my papers and instruments was one, and I was very anxious on that account. Robinson soon returned, not having found the tracks. There was another mount we passed yesterday, two or three miles to the north of us, and I requested Mr. Carmichael to go there and see if he could find water, while Robinson and I found the missing horses, and then we would bring them all on. So for fear any more should retreat in the scrub, we tied the horses up in two mobs to trees on the little plain. I tied one lot and Robinson the other, some little distance apart. We then separately made sweeps round, always returning to the pack-horses the opposite side we started from. We then went again in company, and again on opposite sides singly, but no tracks or horses could be found. I was in a most undesirable state of mind, for five hours had elapsed since I first heard of their absence. I determined to make one more sweep out beyond any I had already taken, so as to include the spot we had camped at, which took me more than an hour; and when I returned I found Robinson had found the two horses in a small but dense bunch of scrub, not twenty yards from the trees he had tied his horses up to. During my last absence he had gone on top of a small stony eminence close by, and from its summit had obtained a bird's-eye view of the scrub below, and thus perceived the two missing animals. On my return I was pleased enough to know that everything was all right, though such a length of time had been wasted, for nearly seven hours had been lost, during which I was riding my horse the whole time at a rather fast pace; it was getting quite late, and I was afraid we should be unable to reach the gully where the water was now, before night, which would be an additional suffering for the horses. We at length pushed on, and Mr. Carmichael joined us in the course of a few miles, he having been unsuccessful in finding any water. As it was getting dusk, we reached the small creek up in the rocks, of which I had seen the water on Sunday. At a certain point the creek split into two, or rather two channels joined and formed one, and I suppose it was by the same ill fate, that had pursued me all day, that I mistook the proper channel, and we drove the unfortunate and limping horses up a wretched, rocky, vile, scrubby, almost impenetrable gully, where there was not a sup of water. On discovering my mistake (for even explorers are liable to these kinds of error), we had to turn them back over the same most horrible places—all rocks, scrub and triodia. At last I got them into the proper channel, not that it was any better than the other, indeed it was worse. When near the place, where I had last seen the water, I dismounted and walked up to see, how it had stood during my absence, and was horrified to discover, that the lowest and largest hole was nearly dry. I bounded up the rocks to the next, and there, by the blessing of Providence, was still a sufficient quantity, as the slow trickling of the water from little basin to little basin had not yet entirely ceased, though its current had sadly diminished since my last visit, only some seventy hours since. By this time it was nearly dark, and totally impossible to get the horses up the gully, so we had to get them over a horrible ridge of broken and jumbled rocks, having to roll away huge boulders to make something like a road to enable the animals to approach the water. Time accomplishes all things, and in time the last animal's thirst was quenched, and the last drop of water was sucked up from every basin, and I was afraid it would not be replenished by morning. We had to encamp in the midst of a thicket of a kind of willow-acacia, with pink bark all in little curls, with a small and pretty leaf; it is of the most tenacious nature, you may bend it, but break it won't; we had to cut away sufficient to make an open square for us to lie down on, and then remove the great bunches of triodia; then, when the stones were cleared, we had something like a place for a camp, and by that time it was nearly midnight, and we retired, all heartily tired with our day's work.

Wednesday, 25th September.—The night was again clear and delightfully cool, the thermometer at daylight standing at 38°. My first thought this morning was to know, how the water-holes looked, and to ascertain, if the slight flow still continued, as, if not, my intended trip to the mountains out westward would be prevented, as I should be compelled to retreat to the eucalyptus-creek, at which we got our last water nearly forty miles away. Mr. Carmichael went up into the rocks to discover, and on his return reported, that they had all been refilled in the night, and that the trickling flow of water continued but slightly less in volume. This was a great relief to my mind. I only trust it will last until I return from those dismal-looking mountains, and enable the pack-horses to remain here in the meantime; I can expect no more. It was necessary to try and discover some more water, if possible; so after breakfast I walked away, but after travelling up gullies and gorges, hills and valleys, I had to return quite unsuccessful, and I can only conclude, that this water was permitted by a kind Providence to remain here in this lonely spot for my especial benefit, for no more rain had fallen here than on any of the other hills in the neighbourhood, nor is this one any higher than others about here, nor is the gully in any way different to dozens of others which I visited, except that this one had a little water and all the rest none. I have in gratitude called this hill Mount Udor, as being the only one in this region, where a drop of that requisite element was to be obtained; and when I left, the udor had departed also. I had to get two of my riding horses shod to-day, as the country I intend to travel over is about half stones and half scrub, and my riding horses are very tender-footed. I have marked a eucalypt in this gully, close to the foot of the rocks where I found the water, $\left[{G \atop 21}\right]$, as this is my twenty-first camp from Chambers' Pillar. My position here is in latitude 23° 14′, and longitude 131°, var. 3° (nearly) E.

Thursday, 26th September.—I was glad to find this morning, that my water supply is still holding out as well as could be expected, the trickling not yet having ceased. The thermometer was high this morning, viz. 54°, and the day gets rather warm, though the thermometer in the shade at three o'clock did not rise higher than 68°, yet as we were in such a confined gully, not a breath of wind could get to us, which I suppose made it feel warmer. In the sun the thermometer stood at 124°. I remained in camp to-day to enable the newly shod horses to get used to their shoes, as one of them was dead lame when shod; there was but poor feed about here, and the horses wandered a good distance away into the scrub, and four of them could not be found.

Friday, 27th September.—The first part of last night was close and sultry, when a slight thunderstorm passed over us; the rain merely sprinkled down for nearly an hour, during which time not sufficient fell to damp a pocket-handkerchief. In the morning the thermometer stood at 60°; the sky was much overcast in many places, and it seems possible that we may have a change. The flies in this gully have been most tormenting for the last two days. The water in the rocks, I am glad to say, is still holding out, and I trust will do so while I am away, and that the party will not have to retreat during my absence. The horses did not arrive very early, and it was long past mid-day when, after watering them, I started away by myself to the ranges out west. I proceeded on our old tracks as far as we went the other day, that is to say, nearly west for twelve miles. At that point some other hills bore in the direction of N. 77° W., which was nearly the line the far ranges lay upon. I reached the foot of the first hills in fifteen miles further. The country alternated between open stones near the hills and dense scrubs a mile away from them. The far range now lay right ahead of me, and bore N. 75° W., with a bed of scrub all the way between. It was near evening when I entered the scrub, which consists of mulga and other bushes. It was the thickest scrub, I think, I ever encountered; it was almost impenetrable, especially in spots, and I had the greatest difficulty to get the horse which I was leading to come on at all, as I had no power over him whatever; he would drag so on my arm, that he nearly pulled me off my horse, and made it most difficult to use the compass, and continue in a straight course. Night at length overtook me in the scrub, and I was compelled to encamp in the midst of it, having travelled thirty-nine miles from camp. During the evening the sky was overcast, and very slight rain fell, and at intervals it sprinkled me during the night. It was a good thing for the horses, as it moistened the dry grass, upon which they were hobbled out. I hoped at the camp they would have had a good shower, as it appeared to be raining in that direction.

Saturday, 28th September.—I was up early enough from my solitary couch, but was compelled to wait until daylight before I could commence to track my horses. The thermometer stood at 60°. The horses wandered back nearly to the last hills we had passed, seven or eight miles, and it was rather late, when I returned to my night's resting-place and saddled up. The sky was still cloudy and the weather warm and sultry; no signs of rain appeared upon the ground. I left this wretched camping-place behind me at last, and continued pushing on through the scrub, and in about seven or eight miles I at length came out of it at the foot of the mountain, for which I had been steering. There was a little creek or gully rather with eucalypts, where I struck it, and I rode up to it. It was as all the others had been, scrubby, rocky and dry. I left the horses and ascended to the top, which was between 900 and 1000 feet above the scrubs which surrounded it. The horizon was broken by low ranges nearly all round, but scrubs intervened between them all. I descended again and walked into dozens of gullies and rocky places, and found some small basins, which would hold water for some weeks; but at the time of my visit they were all dry. I was at this spot nearly eighty miles from a sufficient supply of water; that at the camp, forty-five miles away, would probably be gone by the time I returned, if indeed my companions had not already been compelled to retreat. I could not go any further west on this line under these circumstances, for my horses had already been one night without water. It was now evening again, and they would get none to-night, as I had found none during my walk about the hills, so I left this desolate range, which I called the Ehrenberg-Range, after Professor Ehrenberg, of Berlin, a scientific friend of Baron Mueller, and retreated into the scrub again. I travelled upon a different line this time, going nearly east, hoping to find a rather less thick part of it, but it was all alike. Night again overtook me, and I had to remain again in this scrub, not very far away from where I had slept the night before. I had not travelled very far to-day, only having accomplished about twenty miles, but the day was wasted in an ineffectual search for water round the base of the ranges.

Sunday, 29th September.—I had so short hobbled my horses last night that though the scrub was as thick as possible yet they did not go out of sight, in fact I might as well have tied them up, for they did scarcely go a yard. I continued my east course for seven or eight miles further, which brought me to the north side of a range, round whose southern base I had passed yesterday. I had gone this way in hopes of meeting some small creek or place, where I might find a little water, but none such seemed to exist. I also went round the northern side of the hills (we had been at upon the south side with pack-horses) for the same reason, but no water could be found. It was late in the afternoon, when I reached the camp, and I was gladdened to find the party still there, and that the water supply had held out so long. I had travelled 33 miles to-day.

Monday, 30th September.—The water supply this morning was at a very low ebb, the trickling having ceased from the upper holes, though it was still oozing over the rocks into the lower ones. I intended to pack up and be off from this place, where we have had such misery, for I never was camped in a worse place, I believe. It was an expedition in itself to get water for the camp from the rocks; the horses dreaded to approach it on account of their feet being tender. It required also much work to get sufficient firewood to boil a quart-pot; for, although we were camped in a dense thicket, it was all green and useless for firewood. I intended to retreat from here to-day, but just as Robinson was starting to find the horses a shower of rain came on, and I was in hopes it might end in a heavier fall, so I decided to remain one day longer, especially as the horses would not now (in consequence of the shower, and not being brought to it) touch upon the last remains of our water supply. The trickling had at length ceased altogether, but the little holes were full, and there was just a bare drink for the horses and no more; the rain fell in such slight and gentle showers for two or three hours, that it left no trace of its fall even upon the rocks, so that our water supply was not increased by one pint. I have no alternative but to retreat from this and go back to the creek I came from. Where I left the creek its course was not far from a west point, though from there it seemed to incline more to the south. I thought by going a south course from here to intersect it in the course of 25 to 35 miles, supposing its course carried it so far to the westward of where I left it; so, upon starting from here, I decided to travel upon that bearing. A few quandong trees exist amongst these gullies; also a tree known by the name of the cork-wood tree. The wood certainly is soft and light, but it is by no means a handsome tree. Those I saw were nearly all dead; they grow in the little water channels. The ants here, as in nearly all tropical Australia, build their nests from four to five feet high, to escape I suppose from the torrents of rain, that at times fall in those regions; it also protects their stores and eggs from the fires the natives continually keep going. This fact probably accounts for the absence of insects and reptiles so conspicuous here. One night however I actually saw some glow-worms. The native poplar is also found in the scrubs and water channels of this part of the country. The climate of this region appears very peculiar; scarcely a week passes without thunderstorms and rain; but the latter falls in such small quantities, that it appears useless. I should like to know how much rain would have to fall before any could be discovered upon the surface of the ground. In a salt bush country, with the rains I have met since I started, water could be got at almost every turn in claypans and crabholes; but here no such thing exists, it is all either pure sand or pure rock. The native orange-tree grows here, but the specimens I have met are very stunted; the bloodwood-trees, which always enliven any landscape where they are found, also occur here; but they are not the magnificent vegetable structures which are known in Queensland, but are mostly gnarled and stunted; they also grow near the watercourses.

Tuesday, 1st October.—This morning broke bright and clear, and I was only too thankful that such was the case, to allow me to retreat at last from this miserable encampment, into which the fates had drawn me, as the late sprinkling showers only delayed without benefitting us. Robinson went after the horses, and I went up the rocks to see if yesterday's showers had in any way augmented our water-supply; but I found it was just the opposite, for one of the little holes that had previously watered two horses had not been refilled during the night, the trickling of the water from hole to hole had entirely ceased; had it not been for the showers yesterday, which at all events kept the horses from coming to drink, there would not have been sufficient water for half the horses; but as it was, when Robinson brought them, they did not appear very thirsty, and fortunately there was just sufficient water for them, as some would not drink at all. One mare was away, which Robinson said had foaled, and the foal was too young to walk or move, the dam also was extremely poor, as she had been losing condition for some time previously. I had no other alternative but to send Robinson back, to kill the foal and bring the mare up to the camp, as it was impossible for us to remain any longer at this place; indeed there was not sufficient water for the unfortunate mare when she did come, so I was compelled to be off at once. Mr. Carmichael and I packed up the horses, while Robinson was away upon his unpleasant mission. When he returned, the mare looked the picture of misery. At length I turned my back upon this wretched camp, and away we went to the south, it being half-past two o'clock when we got clear of our prison. I turned to the south in the hopes of again striking the creek, which I had left to the east-south-east, where at my seventeenth camp I had obtained water by digging. As far as I had travelled upon it, its course, as I said before, was very little to the south of west, but it gave indications of having from that point a more southerly trend. But the question was, had it still an existence as far to the west as I now was? That question could only be answered by my going to see. I therefore went and travelled south for twelve miles, at first over rocks and stones, then for two or three miles through thick scrub. The country then became a little less scrubby, and consisted of sand-hills timbered with casuarina and covered with triodia; in ten miles we passed under the lee of a low bluff-hill, and at two miles beyond it we encamped in the sand-hills among casuarinas without water. In the scrub, after leaving the camp, we saw several quandong-trees, and got a few of the ripe fruit. The day had been warm and sultry, though the night set in cool, if not cold. Mr. Carmichael walked to the top of the low bluff, and informed me of the existence of low ridges bounding the horizon in every direction, except to the south-south-eastward, and that the intervening country appeared all sand-hills with casuarinas or mulga-scrub. A yellow flower, which I plucked at the bluff, was an old Darling acquaintance; indeed, I may say, the vegetation of this region resembled in many respects that of the Darling-River.

Wednesday, 2nd October.—The night became cold and dewy, and the thermometer fell to 46°. I was glad of the dew, as it would induce the horses to feed better, they being without water. The creatures however, it appeared, could scarcely have fed at all, for they wandered away in all directions in little mobs of twos and threes, and it was nearly mid-day before we got away from this camp. We kept on for several hours over sandhills and through casuarina-timber in unvarying monotony, when at about four o'clock the little mare, which had foaled, gave in, and would travel no further, consequently I was obliged to leave her in the sandhills. I hobbled her however, as I intended, when I found water, to return with some to her, and in the meantime it was better for her to be hobbled, and thus prevent her from roaming about, while it would save me much trouble upon my return to her. We continued on until we had accomplished thirty-nine miles from our camp in the rocks at Mount Udor, but no signs of a creek or any place, likely to produce water, was visible in any direction; the only difference in the appearance of the country was, that there were less casuarinas now upon it, and that it was more open, though the sandhills and triodia were as before. We passed several quandong-trees in full fruit, of which we ate a great quantity; they were the most palatable I have ever eaten, being sweeter than any I have met before; even when green, they do not taste acid. I also saw a few curragong-trees (*Brachychiton*). At this point I determined to try a little more to the eastward for the creek, and turned away from the line I had been travelling upon, namely S. 25° E., for the cardinal point of east. It was however now past sundown, and by the time we had travelled three miles on that bearing, it was too dark to proceed any further; so we again had to encamp without water, our own supply being so limited, that none of us ventured to eat anything; the day had been hot, and we were rather thirsty at night. Our having eaten so many quandongs was the reason, that we had no great appetite for anything else. The horses had to be short-hobbled, to prevent their straying, and we passed another night in the scrub. We camped under the shade of an enormous curragong-tree.

Thursday, 3rd October.—The night was agreeably cool, though the thermometer did not descend below 54°, and there was no dew-fall. Messrs. Carmichael and Robinson went for the horses, which they secured close by, so that we made an early start from this anything but charming spot. From the top of a high sandhill I perceived, that the eastern horizon was bounded by timbered ridges, and that it was not very probable the creek I was searching for was between me and them. Indeed I had began to conclude, that the existence at all of the creek, at such a distance from where I had left it, was not very probable. The western horizon was bounded by low ridges, which had been continuous for many miles; indeed they had been upon our right from the time we left Mount Udor. I began to think, that the best thing I could do was, to steer away for the point I actually knew the creek to exist at, that is to say, to make straight for my old camp upon it; but I first decided to try a few miles north-east, and if I did not intercept the channel of it in a certain number of miles in that direction, I would turn upon a bearing, that would take me straight back to where I had left it, for I was getting anxious on account of the horses; they had already been two nights without water, and the last water they had was an insufficient quantity; several of them indeed had not drank at all. Having as I said got an early start, I went away to the north-east, intending to travel ten miles upon that bearing, if I did not intersect the creek-channel. At five miles, over sandhills with triodia and scrub, we came upon a mass of eucalypts, which sadly perplexed me, for I could not make out if it was a water channel or not, as the timber grew promiscuously upon the tops of the sandhills as well as in low ground between them; there was no appearance of any flow of water ever having passed by these trees, indeed they looked more like gigantic mallee than eucalypts, only they grew separately; they covered a space of about half a-mile wide, and to the right and left of where we struck upon them they also grew. I thought this might be the extreme point of the exhausted channel of the creek for which I was looking; but Mr. Carmichael and Robinson believed it in no way connected with the creek; however I continued upon my north-east course for two miles further, when I found that some eastern ridges were right before me; they appeared evidently some considerable distance through at the point I should intersect them, and as they were entirely composed of scrub and stones, I did not covet the encounter. They seemed thinner, or a less distance through, a little more to the northward; so as I was fairly tired of looking for this wretched creek, from this point I turned upon the bearing, N. 27° E., which would take me back to my old camp in eighteen miles. It was most distressing to me to have driven the horses all over the country without water, in the hopes of picking up with this wretched creek, which seemed to be the only spot in this desolate region, where water could be found; however I knew where it was, and that I had still eighteen miles to go to reach it, and it was the only thing I could do under the circumstances. I had endeavoured to avoid a direct retreat to that old camp, as there the water could only be obtained by digging; every time the horses had to be watered fresh excavations had to be made, for they knocked the sand in so much. I now turned upon this new bearing straight for the old camp. The first sight I took for the proper bearing went over the ridges at a place, where there appeared a few pines and broken rocks; they were about six miles away. We made directly for them; upon approaching them the rocks appeared upheaved in a most singular manner, and a few eucalyptus-trees were visible at the foot of the ridge. I directed Mr. Carmichael and Robinson to avoid the stones as much as possible, whilst I rode over to see if there was anything like a creek, or place where water might be got by digging. As soon as I approached the rocks (at the base of the ridge) I found that there were several enormous overhanging ledges of sandstone, under which the natives had long and frequently been camped; there was the channel also of a small creek scarcely more than six feet wide. I rode on a bit further to another overhanging ledge and found it formed a verandah, wide enough to make a large cave, upon the walls of which the natives had painted strange devices of snakes, principally white; the children had scratched imperfect shapes of hands with bits of charcoal. The whole length of this cave had frequently been a large encampment; I therefore looked about with some hopes of finding a place, where the natives obtained water from when here, not thinking to find any on the surface, but some little hole, where they obtained it by digging, when I espied about 100 yards away, and on the side of a little valley or glen opposite to the cave where I was standing, a peculiar-looking crevice between two blocks of sandstone; from where I saw it, it was apparently not more than a yard wide, but having the appearance of a receptacle for water. I rode immediately to it, and to my great delight and surprise found a very excellent little rock-tarn of nearly an oblong shape, containing a most welcome and opportune supply of the water I was so anxious to discover. A quantity of green slime was over one portion of the surface, but the rest was all clear and pure water; my horse must have thought me mad, for just as he was preparing to dip his nose to get the drink, he so greatly wanted, I turned him away, and made him gallop off after his and my companions, who were slowly passing away from this liquid prize. When I overtook and hailed them, and held up my hat, they could scarcely believe, that our wants were to be so soon and so agreeably relieved. There was plenty of water for my immediate uses in this little tarn, but the approach was so narrow, that only two horses could water at one time, and we had great difficulty in preventing some of the animals from projecting themselves, loads and all, from the rocks into the inviting fluid. No person, who has not experienced it, can imagine the pleasure which the finding of such a place confers upon the explorer; all his troubles for the time are at an end; thirst is quenched, the horses are unloaded and allowed to roam and graze and drink free from hobbles, and the traveller's other appetite of hunger is at length appeased. How thankful I was for this unlooked for supply of water, instead of travelling some twelve miles further, and then having to dig out the sand from the dry bed of a creek; it was truly a mental and bodily relief. After our hunger had been satisfied, I took a more extensive survey of our surroundings, and found that we were in a really very pretty little spot. Low sandstone hills broken and split into the most extraordinary shapes, forming huge caves and caverns, were to be seen in every direction; little runnels, with a few eucalypts upon them, constituted the creeks. Cypress-pines ornamented the landscape, and a few bloodwood-trees also enlivened the scene—no porcupine-grass, and green grass make up to the explorer at least a really pretty picture. This little spot is indeed an oasis. I had climbed high mountains, and traversed untold miles of scrub, and gone in all directions to endeavour to pick up the channel of a wretched dry creek, when all of a sudden I stumbled upon a perfect little paradise. I find the dimensions of this little tarn are not very large, nor is the quantity of water in it very great, but untouched and in its native state it is certainly a permanent water. It has probably not been filled since last January or February, and it now contains amply sufficient to enable it to last until those months return, as very little evaporation takes place from it; provided of course no such enormous drinkers as horses or bullocks draw upon it, for I do not suppose it would last me much more than a month. I find the water is fifty feet long by eight feet wide and four feet deep. The rocks in which this water lies are more than twenty feet high. The main ridges at the back of it are between two and three hundred feet. The native fig-tree grows here most luxuriantly, and there are several trees close by in full fruit, which is delicious when thoroughly ripe. I had no thoughts of leaving this welcome spot for a few days.

Friday, 4th October.—Last night was cool and breezy, though the thermometer did not indicate below 50°. As soon as the horses were got together this morning, Mr. Carmichael and I loaded a pack-horse with water and started back into the scrub to where we had left the little mare the day before yesterday. I found the spot we left her at bore from this place S. 70° W., and that it would be thirteen and a-half or fourteen miles distant; we therefore travelled upon that bearing, and in thirteen and a-half miles we cut our former track at about a quarter of a mile from where we had left the mare. We very soon picked up her track, and found she had travelled about a mile from where we had left her. We found her standing under an oak-tree truly distressed. Her milk was entirely gone; she was alive certainly, and that is all that could be said for her. She swallowed up the water, we had brought for her, with great gust. I believe she would have drank as much as two or three camels could have carried to her. After her drink she seemed inclined to feed a bit, so we let our other horses out for an hour with her, and then we started away again for the tarn. On this line we did not intersect any of the eucalyptus timber, which we had passed yesterday, though our line to-day was only a few miles more to the north than where we met the trees in question. The mare held up very well until we were close to the camp, when she gave in again; however I could not afford so much trouble with one animal as to be constantly going and coming for her; so I persuaded her to increased exertion, and at last she had her reward by being left standing upon the brink of the water, where she was enabled to drink her fill.

Saturday, 5th October.—Soon after dark last night heavy clouds appeared upon the S.W. horizon, and most vivid lightning played around. The storm seemed to divide in two, and passed away to the north and south of us; but just before daybreak we were awoke with a crash of thunder immediately over our heads; it did not rain however then, but after breakfast we had a few light showers, which were too slight to leave any water even on the flat rocks close to the camp. This appears to me a most extraordinary climate. I do not believe a week ever passes without a shower of rain, but none falls to do any good in the way of leaving water behind. One good day's rain in three or even six months would be infinitely more gratifying to me at least, but I suppose I must take it as I find it; it certainly cools the atmosphere a little, when it does rain. Yesterday, for instance, was a most agreeable day, and I hope to get a few more like it after the last few showers. I found several more caves to-day up in the rocks, and noticed that the natives here have precisely the same method of ornamenting them as the natives of the Barrier-Range and mountains east of the Darling. You see the representation of the human hand here as there upon the walls of the caves; it is generally coloured either red or black; it is done by filling the mouth with charcoal powder, if black (if red, with red ochre powder), damping the wall where the mark is to be left, and placing the palm of the hand against it with the fingers stretched out; the charcoal or ochre powder is then blown against the back of the hand, which is then withdrawn, leaving the space occupied by the hand and fingers clean, while the surrounding wall is all black or red as the case may be. One device here represents a snake going into a hole. The hole is actually in the rock, and the snake is painted on the wall, and the spectator is to suppose, that its head is just inside the hole; the body of the reptile is curled round and round from the tail, though the breadth of the creature is out of all proportion as regards length, being seven or eight inches thick, and only two or three feet long. It is painted with charcoal-ashes, which have been mixed up with emu-fat. Mr. Carmichael has left as ornamentations upon the walls a few choice specimens of the white man's art, which will no doubt help to teach the young native, how to shoot either in one direction or another. It had been raining mostly all day in light and fitful showers; but now it ceased, and it was evident we were not to be benefited by any this time.

Sunday, 6th October.—After yesterday's rains there was a heavy dew upon the grass this morning, but the solar beams very soon caused it to evaporate. Mr. Carmichael and I went out to the eastward this morning, to make sure, if our old creek really came down anywhere in that direction. After crossing the ridges at whose foot we are camped, and which proved about two miles through, we descended into a sandhill and casuarina country, with scrub and porcupine-grass, being the same as that on the other side of it. We travelled for about fourteen miles, until we were so close to a portion of the hills, previously called Gardiner-Range, that I could see with the glasses, that no creek existed between us and it, and we were some twelve miles or so to the south-east of our old camp upon it, thereby proving that it must have ceased to exist at no more than two or three miles from where I left it. We could see the smoke of burning grass to the W.S.W., and still further round to the S., where the natives must have set it alight. I determined to-morrow to visit the country over to the westward of our camp, as there was a continuous ridge for miles, which might have some watering places in it; for, as I before remarked, though the little tarn I had discovered was a most valuable find, yet it would not last me much more than a month. I must therefore endeavour to find some other water, to remove to before long.

Monday, 7th October.—This morning the thermometer stood at 60° at dawn, and it felt as though a warm day was approaching. Mr. Carmichael and I started away to a small rocky eminence, which bore a great resemblance to the rocks immediately behind this camp, and in consequence I thought we might be successful in finding water in its neighbourhood. The rocks in question bore S. 62° W. from camp. We travelled over sandhills, through scrub and triodia, and through some casuarina-country, until we approached the ridge and reached the hill, we had been steering for, in twenty miles. It was composed of a mass of broken red sandstone rock, and was isolated from the main ridge, but others similar were in the vicinity. After searching about we soon found, that there was no water, or any place where water could lodge. We took a round amongst several of the hills on the ridge, and followed one line of them some miles to the west, where it ended in scrub. To the south-west some more hills appeared, some six or seven miles away, and the country between was filled with dense scrub. When we reached this ridge we wandered round it, and searched about, but no water was to be found. It was getting dark when we desisted, and we encamped near to its base, having travelled thirty-nine miles—on various courses. The day had been warm enough; the thermometer in the middle of the day, when we stopped for ten minutes, had stood at 91° in the shade. We saw the smoke of several fires in most directions, mostly round the base of some other ridges to the south-east, which I determined to visit to-morrow. In consequence of seeing so many fires, natives must be about there and water would likely exist somewhere in the neighbourhood; so I determined to give those hills a thorough search, as water anywhere near could only be found amongst them.

Thursday, 8th October.—The night was again warm. Our horses did not wander far, and we made an early start from the encampment. I first mounted the top of the hill, under whose shadow we had slept; the view from there gave very little promise; there were ridges round to the south of this one for a mile or two, then scrub. Further away, nearly S.E., another mass of jumbled hills appeared above the scrub, and here and there some stony hills appeared in it. We searched round the foot of the hills, we were now at, then round the others more south; but no signs of water. We then entered the scrub again, and at eight miles S.E. came out at more hills and ridges. Our search here commenced again. The scrub all round this ridge was burning, as we went through it. Our search here was as unsuccessful as before; so after wasting some time at it, we went away nearly S.S.W. on another ten miles through scrub and over stony hills and gullies. We came to a rather more defined ridge; at least it was a bit longer than most of the others had been; but there was no more water there, than at any other place visited by us. Where those wandering natives get their supply is a mystery to me. From this point another ridge bore away to the N.W.; it seemed entirely isolated, and higher than most of the others. It was a matter of perfect indifference where to go; there were ridges almost everywhere, and I desired to make as much westing as I could. We had searched amongst all these ridges to the E. and S.E. of this line of ridges without success; we had found rocky places, where water would remain for some weeks after being filled, but when such an occurrence had or would take place it was impossible to say. We had wandered amongst such frightful rocks and ungodly places, that I began to think it was useless to search any further for water, but yet the natives were about, burning the grass, and raising fresh fires in all directions; it appeared to me they must get their water from the hollow spouts of some trees, and from the roots of others. We saw a few rock-wallaby and a few bronze-winged pigeons in the course of our peregrinations amongst the rocks. Our horses were getting thirsty, and it was very warm; the thermometer stood at 92° in the shade of a tree. From this point I turned to a ridge to the north-west. We were soon in dense scrub again; here and there we rose on to a sandhill higher than the surrounding scrub; at length at twelve miles we reached the hill. A fruitless search frustrated our efforts here again. To the east rose a more continuous ridge, which we followed under its base, hoping to meet some creek or notice some inviting rocky gully; we saw many of the latter, but none of the former. I continued upon this line until we struck upon our former tracks under the ridge, which lay eight miles south-east of our last night's camp; and, as it was again night, we encamped without water. From this point we had travelled almost in a triangle; the distance we went over, without considering the walking amongst the hills, was forty-three miles, and we were still twenty-nine miles from the tarn—the only water we knew of in this extraordinary region. We had noticed the fresh tracks of natives in one or two places in the burning scrub, but if they saw us, or knew of our presence, they most perseveringly shunned us. The only other sign of their presence in the country besides the continuous fires was an old dilapidated yard, where a very long time ago they had yarded emu or wallaby, though we did not see any old native camps, or such things as old wurlies anywhere, or *mi-mis* or *gunyahs*, or whatever name suits best; I usually call them by the latter. No doubt the natives carry skins with water for long distances, and as they travel along they burn the country right and left; we passed through three different conflagrations on our march to-day. I should greatly like to catch a native; I'd walk him off alongside my horse, until he took me to water.

Wednesday, 9th October.—The horses looked wretched this morning; to be two days and nights in such a country without water, is very severe upon them. We had twenty-nine miles to go, to reach camp, which bore nearly N. 40° E.; and though it was early enough, when we started, it was late in the day, when we arrived at the water, the horses having travelled 110 miles over a country, composed one-half of rock and the other half scrub, without a drink, and this pulled them down greatly. I found the water in the tarn had shrunk; I really don't think it will last more than a fortnight. The day was again warm, the thermometer having stood at 92° in the shade at midday—it was probably three or four degrees higher at three o'clock, but I had not time to wait and look.

Thursday, 10th October.—A day's rest after the fatigue of the last four days is absolutely necessary, before we make a fresh start in some other direction. It is not indeed a day's rest exactly, as I at least have plenty to do, but it is a day's respite from the wretched scrub and a good fill of water. The figs on the largest tree, near the cave opposite, are quite ripe and falling; neither Mr. Carmichael nor Robinson care for them; I eat a good many, but I fancy they are not very wholesome, at least for a white man's digestive organs, as at first they act as an aperient, but subsequently quite the reverse. I have called this charming little oasis "Glen Edith," after my niece, Miss Gill, whose father, my brother-in-law, Mr. George Duff Gill of Melbourne, most generously assisted me with funds for my expedition. I have marked two trees at this camp—one [Giles 24] and another [Glen Edith 24. Oct. 9, 72]; Mr. Carmichael also marked one with his name. The receptacle, in which I found the water, I have called the "Tarn of Auber," after Allan Poe's beautiful lines, in which those names occur, as I thought them appropriate to the spot. He says (if I remember right):—

> "It was in the drear month of October,
> The leaves they were crisped and sere,
> Adown by the dank Tarn of Auber,
> In the misty mid-regions of Weir."

And if these are not the misty mid-regions of Weir I don't know what are. There are two heaps of broken sandstone-rocks, with cypress-pines growing about them, which will always be a landmark for any future travellers, who may seek the wild seclusion of these sequestered caves; the bearing of the water from them is S. 51° W. true, and it is about a mile in that direction from the northern heap, and that with a glance at my map would enable any ordinary bushman to find it; and I will take upon myself to say this much for it, it will support thirty horses for a month at least. I sowed a quantity of vegetable seeds here to-day, also some seeds of the Tasmanian blue gum, some wattles, rye, clover and prairie-grass.

Friday, 11th October.—This morning broke cool and fine, with the promise of a cool day; but the promise became broken, for it turned out quite the opposite. During my last attempt to find water without success I determined, upon my next effort, to try the country more to the south; so Mr. Carmichael and I again mounted our horses—fresh ones this time—leaving Robinson and the little dog Monkey again in charge of glen, camp and tarn, and away we went to the south, over sandhills timbered with the fine *Casuarina Decaisneana* often mentioned already; and I may say, that where they grow it will be almost necessarily a desert. In seven or eight miles we passed some eucalyptus-trees, growing promiscuously on the tops of the sandhills as well as in the hollows, but there was no sign of a water-channel of any kind. At twelve miles we ascended the saddle of a low ridge, which lay athwart our course. The appearance of the country beyond was not a whit more inviting, than that we had just traversed. Upon descending to the lower ground at the southern foot of this scrubby range, we entered upon a bit of better country, thickly covered with green grass, but it was also very thickly timbered with mulga. We saw a few kangaroos and emus, but they were too shy, to get a shot at them. In two or three miles further we left the mulga and entered casuarina-timbered country again, where we noticed the absence of porcupine-grass, except in isolated plots. In four miles further we had another ridge fronting us, and some little way to the left of our course a hill and a hollow or valley, which seemed to offer an easier passage through the ridge than by continuing our course. I turned to it also in hopes of finding some place, where water might be obtained. The head of this valley led us down some very scrubby gullies, thickly set with mallee and mulga, also stony and covered with triodia. The gullies eventually formed into the channel of a small creek, which had some eucalypts on it. It was the first of the kind, met by us for some time, and it raised my hopes of getting water considerably. After following it down a few miles—perhaps four altogether—we saw a place where the sand was damp, and we got some water after scratching with our hands; but the supply was insufficient for the horses, so we proceeded on down the creek and found a small hole with just sufficient to allow our horses to fill themselves, which was indeed a great boon to us; but now, having found a little, we expected a great deal more further down. We found a place, where the natives had recently dug for water, and it appeared as if an abundant supply could be obtained, so I decided to encamp there for the night, greatly pleased with having been so fortunate in our day's travel. The distance we had come was twenty-six miles; courses: first twelve miles south 5° west, and fourteen miles south 15° east. The grass along the banks of this little creek was magnificent, being about eight inches high and beautifully green, the old grass having been burnt some time ago.

It made a most refreshing sight to our triodia-accustomed eyes. The heat had been considerable all day. I tested the thermometer at 12 o'clock, and it then stood at 94° in the shade. The trend of this little opportune creek and the valley, in which it exists, is to the south-east. Here, having found water, of course numerous traces of natives were found also, as we saw old camps and wurleys and some recent tracks on our passage down the creek. I was exceedingly gratified to find this water, as I hoped it would enable me eventually to get out of the wretched bend of sand and scrub, into which I had been led, and which evidently occupies such an enormous extent of territory.

Saturday, 12th October.—Last night was warm and sultry, and the thermometer this morning stood at 62°. Our horses fed all night close at hand; and we were ready to start again to endeavour to discover more waters, only further away, when the usual question arose as to what direction to travel in. West was the point I most desired to make for, and the more towards the west the better; but we first of all followed down the little creek, to see what way it went and if there were any more waters in it. Soon after starting we found the creek meandered through a piece of open plain, which was most splendidly grassed; it was really delightful to gaze upon. The soil was good and firm. There also appeared upon it one or two isolated saltbushes—old friends of mine, and the horses too, it appeared, for they made savage bites at them as we passed. The creek here had worn a deep channel, and in three miles from where we had camped we came upon the top of a high red bank with a very nice little waterhole underneath. It was, relatively speaking, a fine little hole; at all events there was a supply of water there sufficient for 100 horses for more than a month, and an abundance most probably in the sand below. Lower down still we found two or three more ponds, and I believe water is always to be found here, if not upon the surface, at least by digging a few inches below. We followed the creek on for a mile or two further, and found that it soon became exhausted, as casuarina-sandhills and triodia environed the little plain all round, and the short course (of scarcely ten miles) of the little creek became swallowed up in those water-devouring monsters. There was from 6,000 to 10,000 acres of fine grass land on this little plain, and it was such a change from the sterile triodia and sandy country outside it, I could not resist calling it the Vale of Tempe. Overtopping the sandhill-country, which surrounded the creek and plain, and lying still to the south of us, was another ridge timbered with mulga. It was headed by a bluff hill, now somewhat to the south-west of us. We bore away for it, and in ten miles from our last night's camp we entered upon and descended into another valley, somewhat similar to the head of the Vale of Tempe, only it was gullyless, scrubless and sandy, having the casuarinas growing all down it. From a hill I discovered, that more ridges stretched away to the south and south-west far away across the horizon in those directions. To the west the country appeared all broken ridges. I was quite uncertain and undecided as to what point to travel for. At length I determined, as the valley I was now in trended south-west, and as on any other point I should have a lot of stones to get over, to turn my steps in that direction, viz., S.W. In the course of eight miles we had met the usual casuarinas and sandhills, with however very little triodia and not many patches of srub. At eighteen miles we turned the horses out for an hour on a burnt patch, where the grass was green; the thermometer stood at 94° in the shade. Starting again and passing over similar country, we at length reached the ridges we were making for, and found a small gap or pass between two hills, through which we passed; we found it formed into a very small creek or gully, but it gave no signs of having any water, in fact it was too insignificant. It was past sundown when we arrived; and, after vainly looking about for water we camped near the pass, having travelled thirty-four miles from last camp on the following bearings: S. 16° W., 8 miles; S. 55° W., 26 miles. The country on the inside or south side of the pass consisted of a small and well-grassed plain; the grass was as beautifully green here as at the last camp, but we had no water.

Sunday, 13th October.—The night set in very sultry, and no blankets were required for the first few hours; in the morning the thermometer registered 62° as the lowest it had fallen to. The horses rambled a bit in search of water during the night, and we found them up another small gully, which this one runs into; it had a few eucalyptus-trees on its banks, and at one spot we found a place, where the natives had dug for water, but not very recently. We scratched out a lot of sand with our hands, and some water percolated through, but it was too deep to get enough for the horses; we had no means of removing the sand, having forgotten to bring the shovel with us. Upon searching further up the gully we found some good-sized rock holes, but unfortunately they were all dry. We next ascended a hill to view the surrounding country, to endeavour to discover, if there was any feature in any direction, to induce us to visit it in the hopes of finding more water. There were several fires raging in various directions upon the southern horizon, and the whole atmosphere was thick with a smoky haze. After a long and anxious scrutiny through the smoke and haze, far, very far away, a little to the west of south, I descried the outline of a range of hills, and right in the smoke of one fire an exceedingly high and abruptly ending mountain loomed. To the south-eastward other ranges appeared; they seemed to lie nearly north and south. The high mountain was very remote, it must be at least seventy or seventy-five miles away, with nothing apparently between us and it but country similar to that immediately before and behind us, that is to say sandhills and scrub. I was however delighted to perceive any feature, for which to make as a medium point, and which might help to change the character and monotony of the country, over which I have been wandering so long; and I think it not improbable, that some extensive watercourses might be found to proceed from these mountains, which would lead me at last away to the west. For the present not being able to get water at this little glen, although I believe a supply could be obtained with a shovel, I decided to return to the tarn at Glen Edith (which was fifty-five miles away from this spot), and remove the whole camp to the newly found creek, and then return here, open out this waterhole with a shovel, and make a straight line for the newly discovered high mountain to the southward. It was however nearly mid-day by the time I had arrived at these conclusions, as we had walked about in the rocks and gullies, before we discovered the mountain, looking for water during several hours, and as we had thirty-four miles to return to the creek, it took us all the remainder of the day to do so, and it was late when we again encamped upon its friendly banks. The day had been warm enough; the thermometer at three o'clock in the shade had stood at 96°.

Monday, 14th October.—Last night was again sultry, though the thermometer fell lower by 5° or 6° than it did yesterday morning. The morning was cool and pleasant, and we had our old tracks to return upon to the tarn. Alex. Robinson informed me upon our return, that he believed some natives had been prowling around the camp in our absence, as the little dog had been greatly perturbed during two of the nights while we were away. It is very possible that some natives came to the tarn to get water, but as it is so enclosed by rocks they left of course no tracks or traces of their visit.

Tuesday, 15th October.—This morning's was to be the last meal, which we should make at our friendly little tarn, whose opportune waters, ripe figs, miniature mountains and imitation fortresses will long linger in my recollection. Opposite the rocks, in which the water lies, and opposite to the camp also, are a series of small fort-like stony eminences standing apart, which form one side of the glen; the other side is formed by the base of the main-ridge, where the camp and water are situated. It really was a most pleasant little spot, though it had one great nuisance, which is almost inseparable from pines, viz., ants. These wretched creatures used to crawl over everything and everybody, by night as well as by day. The horses took their last drink at the tarn, and we moved away for our new creek to the south. It was late however before we got away, and it was consequently late by the time we had unpacked all the horses at the end of our day's journey, which was twenty-nine miles, as we did not camp until we reached the lowest and best waterholes. The early part of the day was sultry and unpleasant; but towards evening the sky became overcast and cloudy, and the evening set in cold and windy.

Wednesday, 16th October.—I gave the horses a day's rest to-day, and did not leave this camp. I found one horse had staked himself yesterday very severely, and that he was quite lame. I got some wood out of the wound, but I am afraid there is still some left in it. The thermometer at daylight this morning stood higher than usual, viz. 65°, indicating another warm day. The little mare that foaled at Mount Udor, and was such an object of commiseration, has picked up wonderfully since then, and is now in good working condition. I have another mare soon to foal; but as she is nearly fat, I do not anticipate having to destroy her progeny. There were great numbers of bronzewing-pigeons upon this creek, and we shot a considerable number of them to-day.

Thursday, 17th October.—This morning Mr. Carmichael and I again mounted our horses, and taking a week's supply of rations with us started off to the south-westward, intending to visit the high mountains seen a little to the west of south from our last furthest point. We left Alex. Robinson behind in charge of the camp, as he had now got quite used to it, and I did not desire to have him with me. He had my little dog, which when travelling we always carried, for a companion. The little animal is an excellent watch-dog, and not a bird can come near the camp but he will give warning. Alex. had plenty of firearms and ammunition to defend himself with in case of an attack from the natives, which however I did not anticipate; and as remaining at the camp with plenty to eat and plenty to drink admirably suited Robinson, I felt no compunction in leaving him by himself. Upon starting I had intended to proceed on our old tracks to the little glen, from where I had seen the mountains, and where I felt confident we could obtain sufficient water at least for the three horses we had with us, if we took a shovel and opened out the place, where we had obtained some by scratching with our hands; and we might be fortunate enough to find a supply sufficient for all the horses at that little spot, which is thirty-five miles from this camp. Soon after starting however Mr. Carmichael wished to travel upon a south line, which would give us a more easterly route, as he said we had seen other ranges to the east of our former line of march, and that on a new line we might find a better place for water than that in the glen to the south-west, as we already knew that was none of the best; so changing my intention of going to our old encampment, we travelled south for ten miles to a scrubby and stony ridge. We passed after leaving the plain below our camp through casuarina-sandhills and scrub with plenty of porcupine-grass. The view from the ridge was by no means cheering; to the south, a long distance off, appeared the tops of the same line of ridges, which extend up to our old camp to the S.W. One point, a little more conspicuous than the rest, lay 10° to the W. of S. I therefore steered for it, for I must point out to my readers, that only at a ridge can the traveller expect to obtain water, as the country between these formations is mostly a dense bed of scrub, and one must look for a stony or rocky reservoir or else a creek-channel in which to dig, which are indeed few and far between. At the ridge, just reached by us, no signs of such features existed; we therefore had no other line to pursue than to travel on to the next on our course, in hopes that better fortune would attend us. We continued on through the unchanging scrubs and over sandhills, with the casuarinas and the triodias ripping our packbags and clothes, scratching skin off our hands and faces, while the persistent branches continually had to be pushed aside. We reached the hill we were making for at length in twenty miles, and saw at a glance, that no favourable signs of a likelihood of obtaining water were here, as it was merely a pile of stones or rocks standing above the scrub or sandhills which surrounded it. In front of us to the S. the whole horizon was a scrub; to the right of us these low ridges extended far away; to the eastward they ended in a scrubby piece of country, lying between this and another set of ridges; the view indeed was desolate in the extreme. To the S.E. lay another rather higher and distinct ridge, almost a range; it was distant some ten or eleven miles. We had already come thirty miles, and it was late in the day now; however I determined to go over to this new ridge, and after penetrating the usual scrub, which lay between, we reached its base in the dark. When we came to it, there was no sign of a water-channel, only the sloping sides of the stony hills, which form it. It was useless to attempt looking for water in the dark, so we turned the horses out without any, having travelled forty miles.

Friday, 18th October.—At daylight this morning the thermometer stood at 65°, the night having been too warm, in the early part of it, to allow the use of blankets. Mr. Carmichael went after the horses, while I ascended the hill, under which we had encamped. It was precisely similar to dozens of others, previously climbed by me; it was surrounded by scrub, and no watercourse could be seen, nor, though I walked about the hills for over two hours, could I find any rock-holes sufficiently large to retain water for a week. This range was rugged with broken granite, scrubby with mulga and bushes, and clothed in triodia to its summit. To the south a vague and peculiar horizon was visible; it appeared flat, as though a plain of great extent existed there, but as the mirage played upon it, I could not make anything of it. Our old friend the high mountain loomed large and abrupt at a great distance off, and it bore S. 30° W. from here. It was too distant to proceed to it at once without first getting water for our horses, as it was possible that no water existed even in the neighbourhood of such a considerable mountain. I had plenty of time this morning to think over all these and many other subjects, as Mr. Carmichael did not return with the horses until nearly ten o'clock, they having rambled back upon our tracks of yesterday for several miles. By the time they did arrive I had concluded to return at once to the little glen, where we knew of a little water, and which was now some thirty miles away to the W.N.W. or thereabouts; accordingly we started and reached it early in the evening. The day had been hot, and the horses were very thirsty, but they could get no water until we dug a place for them. The thermometer had stood at 98° in the shade at half-past two. Although we had reached our camping-ground, our troubles had not ceased, for we had to set to work and dig out the sand from the little water-channel. We were not long in obtaining a sufficient quantity of water for ourselves, such as it was, being of a thick and muddy and rather nauseous flavour, but the horses had to be tied up to prevent them from jumping in upon our work. We found to our grief, that but a poor supply was to be expected; that though we had not to dig very deep, yet we had to remove an enormous quantity of sand, so as to create a sufficient surface to enable enough to run into the well. We had indeed to dig out a place twenty feet long by six feet wide at the bottom, and six feet deep; at the top, of course, it was much larger. We had to work and wait for about two hours before one of our three horses could obtain a drink. We could let only one down to the water at a time. The water came in so slowly, that it took nearly all the night before the last animal's thirst was assuaged, and by that time the one, which had drunk first, returned for more, as he had got thirsty again.

Saturday, 19th October.—The thermometer stood at 68° this morning, and a warm wind came with the rising sun. I gave the horses a spell here to-day, to allow them to fill themselves, before we made a start for the mountain. We passed the day in enlarging the tank, and were glad to find, that a rather larger proportional quantity of water had drained into it, than we had any reason to expect, there being almost sufficient for one horse to fill himself at one spell. We took a stroll up into the rocks and gullies of those ridges, and found a cave ornamented with the choicest specimens of aboriginal art. The rude figures of snakes were the principal objects, but hands and devices for shields were also conspicuous. One hieroglyph was most striking. It consisted of two Roman numerals, a V and an I placed together, and representing our figure VI. They were both daubed over with spots, and were painted with red ochre. Several large rock holes were seen up in the gullies, but they had all long lain dry. A few cypress-pines grew also upon the rocks in several places. The day was decidedly hot. The thermometer stood at 100° in the shade at three o'clock, and we had to fix up a cloth, to get sufficient shade to sit under. Our only intellectual occupation was the study of a small German map of Australia, given me amongst numerous others by my kind and generous patron, the Baron von Mueller; it is one, published in Gotha by Herr Justus Perthes, and contains all the routes of the explorers. And here I may remark by the way, that the very best maps of this continent, both wholly and in sections, are published by that lithographer, under the direction of the celebrated Professor Petermann of Gotha, whose data have been mainly furnished by the Baron von Mueller, who is well known as our great authority upon Australian geography; and any traveller making any new discoveries should send tracings of them to the Baron, when they will find a place in the next series of Dr. Petermann's published maps. We had only this little map to pore over, and how often we noted the facility with which other and more fortunate explorers dropped upon fine creeks and large rivers. We could only envy them their good fortune, and hope the future had some prizes in store for us also.

Sunday, 20th October.—As our horses had no desire to leave the vicinity of the tank, we were enabled to make an early start this morning considering, that it took nearly three hours before all three horses had drank a sufficiency. The high mountain, I had so long wished to reach, was now our steering point, and it bore from here south 18° west, which allowing 3° for easterly variation made it south 15° west; true, I considered it was distant at least seventy perhaps seventy-five miles. We got clear of the low hills near the glen, and almost immediately entered into thick scrub, varied by high sandhills, with casuarina and triodia upon them. We kept our course for twelve miles, when I noticed the sandhills became denuded of timber, and on our right a small and apparently grassy plain was visible. I took these signs as a favourable indication of a change of country. At three miles further we had a white salt-channel right in front of us with some sheets of water in it. Upon approaching I found it a perfect bog, and the water brine itself. We went round this channel to the left, as it was much wider on the right, and at length found a place firm enough to cross over. We continued upon our course, and on ascending a high sandhill in front of us, I found we had upon our right hand, and stretching away to the west, an enormous salt-expanse, and it appeared as if we had hit exactly upon the eastern edge of it, at which we rejoiced greatly. We continued again upon our course over treeless sandhills for a mile or two, when we discovered we had not quite escaped this feature so easily, for it was now right in our road. It appeared however to be bounded by sandhills a little more to our left, so we turned towards them in that direction, but at each succeeding mile we saw more and more of this feature, it continually pushing us further and further to the east of our course, until at length having travelled some fourteen miles, having it constantly on our right, I found that it swept round under some sandhills, which hid it from us until it lay right athwart our path, as we were travelling nearly south, and this feature here lay east and west. It was most perplexing to me, to be thus confronted by such an obstacle. We walked a long distance on its surface, and to our weight it seemed firm enough, but the instant we tried our horses they almost disappeared out of sight. The surface was dry and encrusted with salt, but brine spurted out at every step the horses took upon it. We dug a well under a sandhill, but only obtained brine. This salt-lake here was apparently six or seven miles across, but whether what we took for the opposite side was an island or the main land I could not say. We could see several islands in its bed, some being very high, and red sandhills, which the mirage caused to appear as though floating in an ocean of water. There were some high red sandhills a mile or two further along the shore to the east, and I thought I might find some place, where the natives obtained water—for sometimes near these salt-depressions brackish water may be found, but here it was not the case. Upon reaching the sandhills in question I could see the lake stretching away to the east or east-south-east as far as the glasses would carry the vision, and it being so frightfully boggy, our intended visit to the mountain was entirely cut off. We made one more attempt to get the horses over it, but they were all three floundering about in the bottomless bed of this dreadful lake before we could look round. I thought they would really disappear from our sight. We were powerless to help them, as we could not get near them for bog, and we sank up to our knees (when once the crust was broken) in hot salt-mud. All I could do was to crack my whip to prevent them from ceasing to exert themselves; and, though it relatively took but a short time that they were in this fix, to me it seemed an eternity. They however at length staggered out of this quagmire—heads, backs and saddles covered with blue mud. The creatures' mouths of course got filled with salt-mud also. They were completely exhausted, when they at length reached firm ground. We let them rest for an hour in the shade of some quandong-trees, which grew in great numbers near the lake, and we gathered a great quantity of the ripe fruit. I cannot say I was much surprised to meet with this feature, as the country I found had been steadily falling. Ridge after ridge, which we had passed, the country at their feet had proved some 100, some 200 feet lower than that further north, so that at the edge of this lake I found the level of it was between 700 and 800 feet lower than the country in the neighbourhood of Mount Udor. I had hoped however to stumble upon something more traversable than this. I had indulged in the hopes of approaching a considerable watercourse or a fresh-water lake, or even a salt-bush country; but this was dreadful, especially so as it prevented me from proceeding to a spot, which appeared so inviting as the mountain, to which I was steering. From this spot, not seeing the possibility of pushing south and thinking that after all it might not be so far round it to the west, I determined to return to where we had first struck upon the salt-channel, and to try what a more westerly line would produce; the salt-channel in question was now some fifteen miles away to the north-westward, and by the time we had reached our tracks at that spot it was dark. We had travelled forty-nine miles. While we rested the horses on the lake under the quandong-trees, the thermometer had stood at 100° in the shade.

Monday, 21st October.—The night was again sultry, and blankets unbearable, though we were of course in the open air. The thermometer at dawn stood at 68°. Having short-hobbled our horses we soon got them this morning. We at first had to get round the brine-ponds, after which we followed along the course of the lake on a bearing of north 63° west, which we continued for seven miles, when we were completely checked by another short arm running up to the north-eastward. We had to get about three miles up it before we could cross, and just before reaching it we had seen some low ridges lying about W.S.W., so after rounding this arm we made for them and reached them in twelve miles. There was no watercourse or channel nor were there any rock-holes in the ridge where we struck it, but for fear of losing any that might exist we circumnavigated one-half of the ridge, as it was formed mostly of two portions, and by the time we had returned to our tracks again we had travelled thirty miles. From the top of this ridge the lake could be seen stretching away to the west or W.S.W. in vast proportions, having several salt-arms running back from it at various intervals. At some considerable distance to the west, or a little to the north of it, was another ridge; but it was too far away for me to travel to it now, as the horses would be already two nights without water, and the probability was we should find none there, though I determined to visit it. But I felt, that I must first return to the tank in the little glen, to refresh our exhausted horses. From the ridge, we were now at, the prospect was indeed wild and weird, with the white bed of the great lake sweeping nearly the whole southern horizon; the country near the lake consisted of open sandhills, thickly bushed and covered with triodia; further from it grew casuarinas and mulga-scrubs. It was long past the middle of the day, when I descended from the hill. I had no other course to pursue, than to return to the only spot, where I knew water was to be had, which was distant twenty-one miles a little to the N.E., so we departed in a straight line for it. I had thus been completely baffled in my endeavour, to reach the high mountain, which I now thought more than ever would be the means of offering a route from this scrubby region into which I had got; it really seemed impossible to get out of it. The mountain in question was named on suggestion of Baron von Mueller Mount Olga, and the great salt-feature, which had obstructed me, was on his request called Lake Amadeus, in honour of two enlightened royal patrons of science. The horses were exceedingly weak; the bogging yesterday had taken a great deal of strength out of them, and the last two days had been hot enough for further weakening the poor animals. The thermometer to-day stood at 101° in the shade. We had twenty-one miles to go yet to reach water, and as the horses were so weak, we went so slow, that it was late at night, when we reached the little tank, having travelled a long fifty miles since morning through the most miserable and disheartening country. There was only sufficient water in the tank for one horse, and the poor creatures had to be tied up and wait their turns to drink, as the water slowly percolated through the sand. By midnight they were all at length satisfied, and went off to feed, and we went off to sleep.

Tuesday, 22nd October.—It was absolutely necessary, to give the horses a day's rest, as they looked very much out of sorts this morning, for they had done then 100 miles without water, at fifty a day, which was rather too severe upon them in this country. We had to erect a shade again for ourselves to sit under, as the day was again very warm. We were so overrun with ants and pestered with flies, that in self-defence we took another walk into the gullies, revisited the aboriginal national gallery of paintings and hieroglyphics, and then returned again to our shade. Again we pored over the little German map, and envied more prosperous explorers. The thermometer had stood at 101° in the shade, and the greatest pleasure we experienced that day was, to see the orb of day descend. The atmosphere had been overcharged with smoke and haze all day; for the natives were ever busy at their grass-fires, especially upon the opposite side of the salt-lake. At night the blaze of nearer ones kept up a more than perpetual twilight, and though the fires were many miles distant their light appeared close. I had also fallen into the custom of the country, and had set fire to several extensive beds of triodia, and they had continued to burn with unabated fury; so brilliant indeed was the illumination around us, that we could have seen to read by the light.

Wednesday, 23rd October.—Last night was again sultry, and blankets utterly useless; the only thing we could congratulate ourselves upon was the absence of mosquitoes. At dawn the thermometer stood at 70°, and a warm breeze gently blew from the north. The horses were got early; but as it took nearly three hours to water them, we did not leave the glen before eight o'clock this time. I intended to return to the ridges, we had last left, and which bore to the west of S.W. some twenty-one miles; but I did not return to the exact spot at which I had previously been, but made for the northern side of a portion of the same ridges, which I had not previously visited, and which were some four or five miles to the north of my former line. I thought perhaps to intercept some little channel or find some rock-reservoir on that side. For the first few miles we travelled over stony ground and low ridges, but all thickly enveloped in scrub. At fifteen miles we came upon a rather extraordinary feature for this part of the country, being no less than a small firm saltbush-plain in the middle of the scrub. It had numerous claypans upon it, but they were all dry. Further on I found the northern face of these ridges just as waterless as the southern. The far ridges to the west, mentioned on Monday, bore N. 85° W. from here, (we had come now twenty-one miles). They appeared higher than the ones we were now at. There was apparently another saltbush-plain upon the line of our travels, which we reached in five miles. It was about three miles long, had many claypans upon it, but all dry. At the far end of it we gave the horses an hour's spell. The thermometer stood at 102° in the shade. We proceeded on, and it was nine o'clock at night when we found ourselves under the shadows of the hills we had been approaching, having them upon the north side of us. I looked about in the dark for a creek channel, but could not find one; so we encamped in a nest of triodia without water, having travelled forty-eight miles, through the usual kind of country, that is to say, partly casuarina-sandhills and triodia, and partly scrubs.

Thursday, 24th October.—Another sultry night made blankets quite unnecessary, the thermometer at daylight standing at 70°, being the lowest it had fallen to. I ascended the nearest hill, and had but one feature to gaze upon, and that was the lake, still stretching away east and west as far as I could discern it. A little beyond where I now was it took a more decided turn up towards the north-west or west-north-west. A further view was obstructed in that direction by an intervening range and some high sandhill-country, which probably formed its southern embankment. Several lateral channels were thrown out from the main-bed at various distances, some broad and some narrow. A line of ridges, with one hill much more prominent than any I had seen about this region lately, appeared close down upon the shores of the lake, and it bore from the hill where I stood S. 68° W., and was some twenty miles off. A long broad salt-arm however ran up at the back of the range between me and it, but it appeared so small immediately opposite me, that I thought it possible to cross it, and so reach the hill on its opposite bank. The hills of the ridge, upon which I stood, were composed entirely of red granite. They formed no creek or water-course, nor could I find any rock-holes amongst them. There was a kind of valley between this and another ridge to the north, but not a eucalypt marked the course of a water-channel. I descended from my survey, and told Mr. Carmichael that there was another ridge with a high hill in it about twenty miles off, which we had better make for; so we departed. In five miles we came to a salt-channel, which was very boggy, but we got over it all right. We travelled another five miles, when we reached the arm, before seen from the top of the ridge. It appeared firm, but unfortunately one of the horses got frightfully bogged, and it was only with the most frantic exertions that we at length got him out. The bottom of this lake or rather arm of the lake (if it has a bottom) seems composed entirely of hot, blue, briny mud. Our exertions made us excessively thirsty. The hill looked more inviting the nearer we got to it; so still hoping to reach it I followed up the arm for seven miles in a north-west direction, but it was quite impassable, and it seemed utterly useless to attempt to reach this range, as we could not know how far we might have to travel, before we could get round this arm. I believe myself it continues in the same direction until it reaches the lake again. I was again in a most undesirable position. I was sixty-five miles away from the only water I knew of, and no likelihood of any nearer. There might be water at the mountain, which I was trying to reach, but it was unapproachable, and I therefore called it by that name. I was also certain enough in my own mind, that had I reached the mount in question my progress further west would still be prevented by the lake, as from the top of the last hills, ascended by us, I had seen the curve of the bald sandhills, which bounded its outer shores, trending round to W.N.W. I could get no water upon its shores, except indeed brine, and I had no appliances to distil that; could I have done so, I would have followed this feature in the hope of some creek or river eventually falling into it; but situated as I was, sixty-five miles from the nearest water, and one hundred miles away from my camp, with only one man left there, I had no choice but to retreat, baffled at all points. From my furthest point a range of hills, evidently a long way off, upon the opposite shore of the lake, bore S. 30° W.; it had a pinkish appearance, and seemed of some length. Mr. Carmichael desired me to name it "McNicol's Range," after a friend of his, which I accordingly did. We turned our unfortunate horses' heads once more in the direction of our little tank, and travelled twenty-eight miles from our furthest point before we camped. Our first course was nearly E. to the opening of the valley, mentioned as having been seen from the hill-top this morning. We travelled down the valley, but found no water-channels of any kind, and we continued on until late in the night, when we struck upon our out-going tracks of yesterday, where we encamped without water, having travelled forty-seven miles, ourselves and horses completely tired out with our day's work. The day had been quite hot enough; the thermometer had stood at 103° in the shade. The heat had been increasing at the rate of 1° per diem for several days. When we camped we were hungry and thirsty, tired, dirty and covered all over with dried salt-mud, and it is not to be wondered at, if we were not in very high spirits, especially as we were making a forced retreat. The night set in close and sultry, and some rain clouds gathered in the sky. At about 1 o'clock the distant rumblings of thunder were heard to the W.N.W., and I was in hopes a storm might visit us—it evidently was approaching; and though the thunder was not remarkably loud, the lightning was most extraordinarily vivid; only a few drops of rain fell, and the remainder of the night was even closer and more sultry than before.

Friday, 25th October.—We were up this morning at earliest dawn, before the stars had disappeared from the sky. The horses having been short-hobbled had not wandered far; they looked indeed most pitiable objects, their flanks drawn in, their eyes hollow and sunken, and as is the case with horses when greatly in want of water the natural vent was distended to an open and extraordinary cavity. Two days of such stages through such a country will thoroughly test the finest horse, that ever stepped. We had thirty-six miles to travel to reach the water, and we could not be sure that the limited supply at that spot might not have fallen off. The horses being so jaded it was late in the afternoon, when they at last crawled into the little glen; the last few miles being over stones made the pace even more slow. Even their knowledge of the near presence of water failed to inspirit them in the least; probably they knew that they would have to wait for hours at the tank, when they did reach it, before their craving for water could be appeased; the day had been hotter than any preceding one; the thermometer standing at 104° in the shade, when we rested for half-an-hour under the shade of a casuarina-tree, and it had not then quite reached its maximum, the thunder and lightning of the previous night not having cooled the atmosphere in the least. The horses had walked 131 miles without a drink, and it was only natural, that the poor creatures should be exhausted. When we arrived at the tank, we had to set to work (as soon as one horse had drained out what water was in it) and throw out the sand which had fallen and been blown into it during our absence. Some natives had also visited the place while we were away, as their tracks were visible in the sand round the tank. They must have stared to see such an enormous piece of excavation. I thought the horses would never be satisfied, and at length when they were so, two of them lay down and groaned and rolled and kicked so, that I thought they were going to die. One was a mare, and she seemed the worst; another, a strong lump of a young horse, which had carried me well; the other was my old favourite riding horse; this time he had only carried a pack, and it was he that had been bogged; he was the only one that did not appear distressed, when filled with water. The other two lay about in evident pain till morning. About the middle of the night thunder was again heard, and flash after flash of even more vivid lightning than that of the previous night was again seen; indeed they were so vivid, being alternately flashes of fork and sheet, that for nearly one hour the light never ceased. The thunder was also much louder than last night's, and a slight mizzling rain fell for an hour. The barometer had fallen considerably for the last two days, so I expected a change. The rain which fell was too slight to be of any use. The temperature of the atmosphere however was considerably changed, and by the morning the thermometer had fallen to 48°.

Saturday, 26th October.—This morning was most agreeably cool after the storm. As our horses were not fit to travel, I gave them another day's rest, for they looked most wretched. They were still craving for water, and were round about the tank the whole night. It was not very pleasant having to remain here, with nothing to do but consult the little map again and lay off my positions on it. My furthest point I found to be in latitude 24° 38′ and longitude 130°. It was the second time I had reached nearly the same meridian, for my former furthest western point was only a little to the east of north from this point, though it was 100 miles away. I had been repulsed at both points—in the first instance by dry sandstone-ranges, in the midst of dense scrubs, without water; and in the second by a salt-lake equally destitute of fresh water. It appears to me plain enough, that a much more northerly or a more southerly course must be pursued, to reach the western coast; at all events in such a country it will be only by time and perseverance, that any explorer can penetrate it. I think I remarked before, that we entered this little glen—through a pass—about half a mile long, between two hills of red sandstone. I named this pass after another friend of Mr. Carmichael, and called it "Worrill's Pass." The little glen, in which we dug out the tank, I could only call "Glen Thirsty," for whenever we returned to it, ourselves and horses were choking for water. Our supply of rations, although we had eked them out with the greatest possible economy, was gone, for we brought only a week's supply, and we had now been absent ten days from home; we had also one day's more starvation after to-day before we could reach it. However, as the horses were unable to travel, we had to remain *per force.* The day became again warm enough, as the thermometer at 3 o'clock stood at 100° in the shade, but as there was a cool breeze blowing the temperature seemed agreeable. During the day I had a long conversation with Mr. Carmichael in regard to our affairs in general and our stock of provisions in particular, and the conclusion we arrived at was, that we had been already nearly three months out, and had not progressed so far in that time as we had expected. We had found the country exceedingly dry, and until some rain fell it seemed scarcely probable, that we should be able to penetrate further to the west; and if we had to remain in depôt for a month or two, it was necessary by some means to economise our stores, and the only way to do so was to dispense with the services of Alex. Robinson, whom we had left behind us at the camp. It would be necessary in the first place of course, to find some creek to the eastward, which would take him to the River Finke; and by the means of the same river Mr. Carmichael and I might eventually get round to the southern shores of Lake Amadeus and reach Mount Olga at last. In our passage up the Finke-River two or three creeks had joined it from the west; and as we were now beyond the sources of any of these, it would be necessary to discover some road, to one or the other of them, before Robinson could be sent away. By dispensing with his services we should then have sufficient provisions to enable us to hold out for some months longer; and if we had to wait so long as the usual wet season in this part of the country, we should still have several months' provisions to start on again with. To all these arguments Mr. Carmichael fully agreed, and it was decided, that I should inform Alex. of our resolution as soon as we returned to the camp, where we had left him.

Sunday, 27th October.—This morning broke with the promise of a cool day. We got our horses early, and after the usual three hours' (nearly) suction at the tank, we turned our backs for the last time upon this little spot, to which we had so often returned with exhausted and choking horses, not having met with another place in the whole of our travels in this part of the country, where another drop of water was to be found. Having been so much longer away than I anticipated, I was getting anxious to learn how matters were progressing at the camp with Robinson. We had thirty-five miles to go upon a nearly N.E. course to reach the camp. As we were travelling through Worrill's Pass (a mile from the tank) we noticed that scarcely a tree in it had not been struck by lightning. The boughs were lying in all directions in the pass, and several cypress-pines on the slopes of the mountains on either side of it had also been struck down. I thought indeed at the time, that it would be strange, if such a storm as we had experienced two nights previously did not leave some marks of its fury in our neighbourhood, for only twice in my life had I seen anything like it before—once upon the Tumut in New South Wales, and once in the old country. We reached the camp again at 5 p.m.; my old horse, which carried the pack, having gone quite lame to-day, and caused us to travel rather slowly. Robinson was alive and well, and my little dog Monkey was quite overjoyed at our return. Robinson reported, that natives had been frequently in the neighbourhood, and had lit fires within a few hundred yards of the camp, but they had not actually shown themselves. The mare, which had been heavy in foal when we left, had been safely delivered of a daughter; but the horse, which had been staked and was so lame when we left, was now far worse, as he could not put his foot to the ground; there was evidently some wood still in the wound; and I found also, that my old horse, which had gone lame with us, had run a stake into his coronet. I probed the wounds of both of them, but could not get any wood out.

Monday, 28th October.—To-day I remained in the camp, as indeed both Mr. Carmichael and myself were entitled to a day or two's rest after our late travels. I also wanted a brief respite to consider, which way to bend my steps. I had been repulsed to the north-west by dry mountains and dense scrubs; and to the west was a long broad impassable salt lake; and to the south and south-east the same feature obstructed me. I had decided to send away Robinson, as by that means our ration-supply would hold out much longer, and when any rain fell (if we had to wait for it) Mr. Carmichael and I could get on well enough without him. The lame horses I find are worse; I cannot find any wood in the wounds, but I feel sure there must be some. It is a great pleasure to return to a place where sufficient water can be obtained for a week.

Tuesday, 29th October.—I remained again at this camp to-day, having told Robinson I should not require him any longer, after I had found a road for him to reach the Finke by, and on

Wednesday, 30th October, Mr. Carmichael and I having got three fresh horses started away again from the little creek, upon which our camp had so long stood. To the eastward of this place (in my travels to the south-west) I had noticed a conspicuous range of some elevation, quite distinct from the ridges, on which we were now camped, and having an almost overhanging crag, forming its north-western face. To get out of the ranges however, in which our little creek exists, we had to follow the trend of a valley, formed by a succession of what are known as reap-hook hills, which ran about E.S.E. In four and a-half miles we crossed a very small and insignificant little creek, with a few eucalyptus-trees on its bank; it however had a small pool of water in its bed, where we struck it, and at which we watered our horses as we passed. We followed down the valley, having the hills on our left hand, that is to say on our north side; the valley was well grassed and open in most places, and the triodia was fortunately absent. At ten miles from the camp we turned up through a small pass a little to the north of our former line, and entered another valley, which ran parallel to the last, and in which I noticed several peculiar conical hills. We passed over a saddle between two of them, and then found a flat open valley running all the way to the foot of the higher range I was steering for. It appeared very red and rocky, being composed of enormous masses of red sandstone, the upper portions of it being quite bare, with the exception of a few cypress-pines moored in the rifted rock, and I suppose proof to the tempest's shock. On nearing the mountain I saw, that a creek, lined with eucalyptus, issued from a gorge, and it had some fine-looking timber upon its banks. Having reached the bank we followed it up towards the foot of the hill, for a mile or so, and Mr. Carmichael found a fine little sheet of water in a stony hole; it was about 400 yards long, forty yards wide, and had about four feet of water in it. The grass here was green, and all round the foot of the range the country open, well grassed, and very agreeable to look at. We had travelled eighteen miles, and having found so pleasant and eligible a spot, we went no further for the day. We took a stroll up through the gorge into the rocks, and found several rock-waterholes with plenty of water in them. I also noticed a few stunted palmlike zamias endeavouring to eke out a scanty subsistence amongst the rocks. On returning to the camp from the side of the mountain, I saw that the creek trended to the south-west, and passed through another ridge two or three miles away; there was some smoke there and no doubt natives were about, as most likely a large body of water lodges there, the creek running under the lee of a precipitous red sandstone hill and through a glen. The weather to-day was most agreeable, and since the heavy thunderstorms, experienced at Glen Thirsty, the temperature has been much cooler; the thermometer to-day did not rise higher than 99° in the shade. Before reaching this creek, we crossed the head of another running round the north-western crag, mentioned before. I called it "Johnny's Creek," after Mr. John McCulloch, of Gottlieb-Wells and Black Rock, who showed me great hospitality and kindness when I travelled past their stations. The creek, upon which I encamped, I called "King's Creek," after Mr. Fielder King, who is a resident with Mr. McCulloch, and also an old and kind friend of mine. I called the crag round which Johnny's Creek turns, "Carmichael's Crag," after Mr. Carmichael.

Thursday, 31st October.—We got an early start this morning, as the horses fed close to the camp all night. The thermometer at daylight stood at 65°, which indicated we might expect a rather warmer day than yesterday. As I had in view the finding of some tributary, which would run into the River Finke, I determined to continue along the southern face of this range, which indeed was the only line practicable just about here, and having found one creek issuing from it, I hoped to be equally successful further on, and find one running from this range into the Finke-River. On leaving the camp we followed close along the foot of the range, and had an open grassy country before us, with the range forming a high wall on our left. It was the most agreeable and pleasant piece of country, which I had met in my travels in this part of the world; our course along the foot of the ranges was nearly south 60° east. In four and a-half miles we came to the bank of another eucalyptus-creek, coming immediately out of the range; we followed it up a short distance, and found it took its source from the drainage of the range, which issuing under a high and precipitous rock-wall, formed a small swamp, thickly set with reeds and having plenty of water amongst them. The reeds covered a space of several acres. I called this "Penny's Creek," after a Mr. Penny of Yorke's Peninsula, who had been desirous of joining my expedition. I believe the water permanent. It runs from under the rock, but it does not show in the creek-channel below the reedbeds. There was a similar reedbed and creek about half a mile beyond this one, but I did not go up to it, but crossed its channel where it was dry, and continued on my course. At eight miles from last camp the line of range trended up a little, and our course was now nearly S. 70° E. On this line we travelled seven miles, and crossed the channels of two other eucalyptus-creeks, which had no water, when we crossed them; but no doubt up under the hills there was plenty. At eighteen miles from the camp we arrived upon the banks of a creek, much larger than any we had previously crossed; and as its trend was E.S.E. I thought it might possibly be large enough, to reach the River Finke. Its bed was dry, where we struck it. I turned up it towards the range, and followed it up to a gorge or glen rather, as it was more open than a gorge; here we found three fine rock-waterholes, all having good supplies of water in them. The glen and rock is all red sandstone. One of these holes reminded me of the waterhole at Sturt's Depôt Glen in the Grey-range, only the rock formation is entirely different. At Sturt's Depôt they are composed of granite and slate in vast sheets or slabs standing in rows, and almost perpendicular; here the rock is red sandstone, and though a cliff overhangs both places, and there are other points of resemblance, yet the water here is not in such a body as at the veteran explorer's camp. I have named this "Stokes's Creek," after Frank Stokes, Esq., of Coonatto in the North. We turned our horses out here for an hour, and had a swim in the rocky basin, which was most exhilarating; and I could not but reflect, how different this line of country was to that further to the west, where we could not get water to drink, much less to swim in. The day was sufficiently warm, as the thermometer stood at 102° in the shade, when we resaddled our horses and started again. It was rather late in the afternoon, when we did so; for we had lost some little time in poking about the creeks; and time no more stands still with the explorer in the wilderness, than with the banker, the merchant or the broker in the city. From here the hills trended again nearly S. 60° E., and we travelled upon that bearing, still under the foot of the range, and having still a fine stretch of open grassy country to travel over. In five miles after starting again we came upon the banks of another creek. I turned up the channel and soon came to water, and another extensive reedbed filling up the whole glen. The water had a slightly disagreeable vegetable taste, owing to its stagnation. I called this "Bagot's Creek," after Mr. John Bagot, of the Peake, who showed me great kindness, when passing his station. From this creek the line of range trended about S. 75° E., and for some few miles I could discover, that no other creek issued from it, but in hopes of getting other waters we continued on for thirteen or fourteen miles, and though we had passed so much water in the day, yet, such is often the explorer's fate, we got none at night, and it was dark when we encamped, having travelled thirty-five miles. The feed was green where we turned out, and as the horses had so recently drank, they could not be very much distressed at not having more water now.

Friday, 1st November.—Last night proved exceedingly cold; so much so, that the insufficiency of my blankets compelled me to make a large fire, to enable me to pass the night in comfort, and at daylight the thermometer was down to 45°. We got our horses early, and proceeded still along the foot of the range, which here ran nearly east. We also had a low ridge now upon our right hand, and were thus in a valley, which became somewhat scrubby. At five miles through the scrubby valley we came to a place, where there was a sudden fall into a lower valley, which was quite open; and here I discovered, that the range we had so long followed ended at last, but sent out one more creek, which we could see on the flat beyond. Still further to the east another and entirely different range of hills occupied the whole horizon, and one high bluff in it bore nearly east. We descended into the lower valley, and got on to the creek, which I found to be a clay-creek, having some very nice little clayholes in it, but these were dry; at length however at about eight miles from where we had camped we found water in it, and our horses got a good drink. The creek appeared to be making straight for the other range, and I thought, if it actually went through that range, it must eventually reach the River Finke; we therefore followed it down further, and in three miles from where we had got the water found it emptied itself into a much larger, stony, mountain-stream. I called this clay-creek "Trickett's Creek," after a friend of Mr. Carmichael's, who wished to commemorate him. The range, which we had so long followed, which sends out so many watered creeks, and which is over forty miles in length, I called "George Gill's Range," after my brother-in-law, Mr. G. D. Gill, of Melbourne. The country round its foot is the best, which I have seen in this region; and no doubt before many years pass it will be part or all of a stocked run. The new creek, just discovered, had a large stony waterhole immediately above and below the junction of Trickett's Creek; and, as we approached the one below the junction, I noticed three or four native wurleys just deserted by their owners, who, having seen us while we never thought of them, had fled at our approach and left all their valuables behind them, which consisted of spears, clubs, shields, drinking vessels and yam sticks, with other and all the usual appliances of a well-furnished aboriginal gentleman's establishment. Three young native-dog puppies came out however, to welcome us; but when we had dismounted and they smelt us, not being used to such refined odours as our garments probably exhaled, they fled also. The natives had left some food cooking; and when I cooeyed they answered, but would not come to us. This creek was of some size, and its waters are permanent. It seemed to pass through this new range in the direction of the bluff-hill, mentioned before as to the east of us from Trickett's Creek. I named this hill "Mount Levi," and the range of which it is a part "Levi's Range,"—after the well-known Philip Levi, Esq., of Adelaide, who showed me great kindness, when last there. The creek came from the north-westward, apparently draining the country between this and Gill's Range. I named this watercourse "Petermann's Creek," after the celebrated geographer, Professor Dr. Augustus Petermann, of Gotha. We were now sixty-four miles from the main-camp, and had been most successful in our efforts, to find a route to allow of letting Alex. Robinson depart; for there seemed no doubt, that this creek at all events would reach the Finke-River (though I found afterwards it didn't); for in my passage up that creek I had noticed junctions of creeks from the west at my second, fifth and sixth camps, but which one this would prove to be, I could only guess (and my guesses did prove quite wrong); but at all events the River Finke itself was now not more than sixty-five miles further to the eastward. I intended upon returning here, and after despatching Robinson, to endeavour to discover some line of country round the south-east extremity of Amadeus-Lake, and reach Mount Olga at last. Seeing that this creek ran through hills for some miles, and thinking it would gather strength from them on its course, I did not follow it down any further eastward, intending to leave that honour for Alex. Robinson. We turned our horses' heads again for our camp home, and continued travelling (having given the horses an hour's feed in the afternoon) until we reached Stokes's Creek again, and there we encamped at the large rock-waterhole, having travelled forty miles.

Saturday, 2nd November.—We made an early start this morning, and intended reaching the depôt by evening. As we were approaching Penny's Creek, we saw two natives looking most intently at our outgoing horse-tracks, and along which they were slowly walking with their backs turned to us. They did not see or hear us until we were close upon their heels. They each carried two enormously long spears—two-thirds wood and one-third reed; they had also an implement, with which they project these spears, having a kangaroo's or wild dog's claw fixed at one end. They also had small narrow shields. Their hair was tied up in a kind of chignon at the back of the head, being dragged back off the forehead from infancy, a custom which prevails among the natives of this part of the interior. It gives them a wild though sometimes effeminate look. Some wear their hair in long thick curls, reaching down to the shoulders, and beautifully cultivated with iguana's fat and red ochre. This applies only to the men; the women wear their hair cut short. As soon as these two first heard and then looking round saw us, they went off like emus, running along as close to the ground as it is possible for any two-legged creature to do. One of these was quite a young fellow, the other full-grown. They ran up the side of the hills, and kept travelling along parallel with us; but, though we stopped and called and signalled to them, they would not come close; and when I attempted to go to them they ran away. They continued alongside us until we reached King's Creek, where we turned our horses out for an hour. The day was most charming; the thermometer did not rise higher than 89° in the shade. Soon after reaching the creek we became aware, that several other natives were in our vicinity—hearing our original two shouting and yelling—until presently we saw a whole nation of natives coming from the glen, where I noticed this creek runs through, and where I saw smoke on my first arrival at this creek. They commenced shouting, yelling and gesticulating in the most furious manner, our former two now approaching much nearer to us, and when I called them they came within about twenty yards of me, holding their spears in such a position, that they could use them if required. Neither of them were very handsome or very fat; but they were very well made, and as is the case with most of the natives hereabouts, were rather tall—viz., five feet eight or nine inches. When they had come close enough, as they thought, one of them—the elder—commenced to harangue us, and evidently desired us to know, that we were to be off forthwith, for he began to wave us away in the direction we had come from. The whole tribe then took up the signal, and talked and yelled, waving their hands and arms at us. They however did not appear to me to meditate any attack, so I took no notice of them. When it suited us to depart we left them, and were of course saluted with yells and curses in their charming dialects, until we were fairly out of sight and hearing. We reached the camp again by the evening, having travelled thirty-seven miles. No natives had been seen by Alex. at the camp, and nothing had occurred in our absence.

Sunday, 3rd November.—This morning I inspected the two lame horses again, and found they had got worse. I could not leave them here, as the water was not sufficiently permanent. One of them is my poor old favourite, and I would not like to eat him, and the other is too poor to be converted into meat. His lameness, and the pain he suffers, has made him lose what little condition he had at first. It being Sunday I did not shift the camp. About the middle of the day we had a very sudden thunderstorm, but very little rain; the remainder of the day was quite cool.

Monday, 4th November.—To-day, I requested Mr. Carmichael and Alex. to take the two lame horses over to the little five-mile creek we had crossed on our journey to the eastward, so that to-morrow, as we passed by with the other horses, we could take them on with us, as they would be all the fresher. When they returned they said they had great difficulty in getting them along, and they are afraid they cannot travel at all from there. I thought, if I could get them to King's Creek, to leave them there.

Tuesday, 5th November.—I broke up this little depôt camp at last, having had the camp there since the 15th of last month; many delays had occurred to me, and prevented me reaching a much more distant point, as I so ardently wished. I never expected, after being nearly three months out, that I should be pushing to the eastward, when every hope and wish I had, was to go in exactly the opposite direction; but the explorer must go as the country permits, and I could only console myself with the thought, that I was only going to the east to be enabled at last to go west. I have great hopes, that when I can once set my foot upon Mount Olga, also my route to the west will be unimpeded. Having been away so much from this camp, I have not seen all the horses for some time, and when they were mustered this morning, I found that they had all greatly improved in condition, and almost the fattest animal amongst them was the little mare, which had foaled at Mount Udor, and had been left knocked up in the scrub on the other side of Glen Edith; the other mare, which had lately foaled, looked very well also, and the horses in general were in very good condition. It was past twelve o'clock by the time we turned our backs upon this little creek. At the little five-mile creek, we found the two lame horses, and proceeded on to King's Creek, where we arrived somewhat late in the afternoon. As we neared the creek, we saw several native fires, and immediately the whole region seemed alive with natives—men, women and children, running from the highest points of the mountain down to join the tribe below, eventually all joining in one mass. The yelling, howling and gesticulating, kept up by them was most annoying. When we commenced to unpack the horses, they crowded round us, carrying their long spears, and knotted sticks, and other fighting implements, though I did not notice any boomerangs amongst their weapons. They were growing troublesome, so I rode towards them, thinking they would as usual retire; but not so—they were in sufficient number to be a bit impudent. I cracked my whip at them—they all made a sudden pause, with a sudden shout, still they did not seem at all inclined to depart. I then unstrapped my gun from off my saddle; holding it up I warned them away, and to my great astonishment away they all went, and we neither saw nor heard any more of them, either that night or the next morning. I came to the conclusion, that they had heard from some other half-civilized natives, who had visited the telegraph line, that unless a white man had a gun, he was no more dangerous than an emu or a kangaroo; so they merely wanted to find out, if any of us had such weapons, and when satisfied on that point, and knew we were not to be trifled with, they very quietly went away, and gave us up as a bad lot.

Wednesday, 6th November.—It seemed to me, that the travelling had rather improved the go of the lame horses, and I decided to bring them on as far as they would go, as if I left them here the natives would no doubt spear them. We made a start away from here and reached Stokes's Creek, where we encamped at the Big Rock Waterhole. As we had started early, we got to the camp early also; and after a frugal dinner occurred a circumstance, which completely put an end to my expedition; for here Mr. Carmichael informed me, that he had made up his mind, not to continue in the field any longer, for as Alex. Robinson was going away, he should do the same. The reader must understand, that Mr. Carmichael was a volunteer, and had his own horses, and was possessed of some little property, and I had no control whatever over his actions. He had accompanied me from Melbourne. We had never any fall out, and I thought he was as ardent in the cause of exploration as I was myself, so that when he informed me of his resolve, it came with greater force upon me. My arguments and entreaties were alike in vain; in vain I showed how, with the stock of provisions we had, we might keep the field for months. I offered to retreat to the Finke even, and follow that river, so that we should not have such arduous work for want of water, but it was in vain.

Thursday, 7th November.—It was with distress, that I turned into my blankets last night after what Mr. Carmichael had communicated to me, and it was with an equal sadness, that I arose this morning; I scarcely knew what to do. I had a lot of horses yet, heavily loaded with provisions; but to take them exploring by myself was impossible. We only went a short distance to-day, viz., five miles, to Bagot's Creek. I here renewed my arguments and entreaties to Mr. Carmichael, not to leave. His reply was, that he had made up his mind, and nothing should alter it. The consequence was, with one man I had discharged, and a companion who had discharged himself, any further exploration in such company was impossible. I had no other object now in view but to hasten my return to civilization, in hopes of reorganising my expedition. And the reader may guess my surprise, when the first news I heard upon regaining the Charlotte-Waters Station, was to the effect, that during my absence two other expeditions had been started out, to attempt the same undertaking, from which I was returning a baffled and beaten man. I was now however in full retreat for the telegraph line; and as I still traversed a region previously unexplored, I may as well continue my narrative to the close.

Friday, 8th November.—To-day the party now silent still moved under my directions, as I had yet to bring it to the telegraph-line. We travelled over the same ground, Mr. Carmichael and I formerly did upon the 1st instant, that is to say, we went from Bagot's Creek to Trickett's Creek, twenty-one miles, and encamped at the first water we had found in that creek.

Saturday, 9th November.—This morning, at starting, we followed this creek to its junction with the River Petermann in the Levi-Range. The natives and their dogs had departed from their old camping-ground, and the place knew them no more. The hills on both sides of this creek approached it near enough to form a valley, consisting of very good country: there were numerous waterholes at short distances all down its bed, as far as the valley continued, from where we first entered it. After running east for a mile it curved round under Levi's-Range to the east-north-east, and at twelve or thirteen miles from our camp, on Trickett's-Creek, the hills of Levi's-Range fell off, and ran down southward, leaving the country open in that direction; the hills on the northern side trending more south-easterly, forced the creek into that direction. From where Levi's-Range fell off no more water was found in the creek, although the bed was broad and the timber vigorous-looking. After travelling ten miles, from where the valley opened out, the creek ran more directly to the eastward, and at that distance I saw a few eucalyptus-trees immediately under some hills two or three miles to the right; and as some smoke appeared above the hills, I knew, that the natives must have been there lately, and that there possibly might be water, so let Mr. Carmichael and Robinson go on with the horses, while I rode over. I found it was the valley of a small eucalyptus-creek, which narrowed into a glen the further I penetrated. The grass on all the hill sides was burning, and as I went further up I could hear the voices of the natives. I felt pretty sure of finding water, so on I went. I soon saw a native in front of me, leisurely walking along with an iguana in his hand, taking it home for his supper. He carried a spear and shield, and had long locks hanging down to his shoulders. My horse's nose nearly touched his back, before he was aware of my presence, when suddenly he gave a start, held up his two arms, dropped his iguana, and gave a most tremendous yell as a warning to his tribe, and bounded up the rocks in front of me like a wallaby. I then passed an eucalyptus-tree, in whose foliage two aged warriors had secreted themselves. I stopped under the tree and looked at them; they certainly presented a most ludicrous appearance. A little further on there were rows of wurlies, and I could perceive the men urging the women and children away, whilst they snatched up their spears and what other weapons they could hastily man, and then they took to the rocks. The glen here was very narrow—indeed almost a gorge, and the rocks were not more than 80 to 100 feet high. It is no exaggeration to say, that the summits of the rocks on either side of the glen were lined with natives; they could almost touch me with their spears if they had wished to do so. It must not be supposed I felt quite at home in this charming retreat, although I was the cynosure of a myriad of eyes. The natives stood upon the rocks like statues, some holding their spears menacingly towards me, and I certainly expected, that some dozens would be thrown at me. In the channel of the creek the natives had dug out the sand in several places for water, and at the bottom of one or two holes, that I looked into, I saw water, but without a shovel only a native could get a drink out of such a small place. I was very thirsty and dismounted, and picked up a small wooden utensil from one of the wurlies, and managed to get a few thimblesfull of dirty water out of the hole. There were only these few small places, as far as I could see, where all these natives got their water. At all events, as the grass was all burnt, and not a blade within a couple of miles, I did not think it a very desirable spot to bring the horses to, especially as by the time of my rejoining the party they would be some miles down the main creek. As soon as I turned my horse round, and he made a few steps towards the way I had come, such demoniacal yells greeted my ears as I have certainly never heard before, and do not wish to hear again. The echoes of the voices of these, I suppose, indignant and infuriated creatures reverberating through the defiles of the hills, and the uncouth sounds of the voices themselves smote so discordantly upon my own and my horse's ears, that it is no untruth to say, we went out of the glen faster—much faster—than we went in. I heard a strange sound as of sticks or other weapons, striking upon the ground behind me, but I did not lose time or ground by looking round to see what caused it. Upon rejoining my companions I merely told them, that I had seen water and natives, but that it was not worth while to turn back, as there was no supply of the former for the horses. Where I overtook them again the creek had ran into several separate channels, and in two or three miles further I found, that it completely exhausted itself upon the plain. I decided still to continue on an east course, as we must intercept the Finke in the course of forty-five to fifty miles upon that bearing; it was however near night, and getting dusk, when we desisted from further travelling, camping without water under the low hills, which ran along to the north of us. We had come a long distance since morning, namely, thirty-five miles. The further the lame horses were driven, the better they seemed to go, and I hoped we might possibly travel their lameness out of them, as I have known instances of that nature occur with me three or four times.

Sunday, 10th November.—Having camped without water, the horses were all short-hobbled, and we were under way again by six o'clock. I intended now to make as easterly a course as the configuration of the ranges to eastward and north of me would permit, so as to strike the Finke in as short a distance as possible. We had however scarcely proceeded two miles, when we struck the bank of a broad sandy bedded creek, which was almost as broad as the Finke itself; its bed was dry when we came upon it, and its course from out of the hills, into which I could trace it for some distance, seemed to be from the north-west. It was however running almost east when we came upon it, but turned again southwards, and it was evident its general trend was south-east. We followed along its bank for about three miles; and Robinson, who went down the bed to look for water, found two large ponds under a hill, which ended abruptly at the creek: there was also a low ridge on the opposite bank, but it did not quite reach the creek; the two however formed a pass, through which the creek ran. I did not go further than these ponds, as the horses had a long day of it yesterday. In the afternoon Mr. Carmichael and Robinson caught a great quantity of fish from these ponds with hook and line. These ponds contain a considerable body of fresh water, and are deep and permanent. I called this place "Middleton's Pass" and "Fish Ponds," after my old and kind friend upon the Darling-River, A. D. Middleton, Esq., of Mena-Murty. The country near this creek was open and grassy, and fit for stock.

Monday, 11th November.—I did not remove the camp from these ponds to-day. Messrs. Carmichael and Robinson caught again great quantities of fish; they were a species of perch, though the largest one, that they caught, weighing I daresay three pounds, had a great resemblance to Murray-cod.

Tuesday, 12th November.—We left Middleton's Ponds this morning, and started away on a S.E. course, as from the top of the hill at Middleton's Pass I saw the creek ran in that direction. In six miles we crossed it, and travelled on its western side. At two or three miles further I found the creek had taken a more easterly turn, and we consequently had to turn in that direction also. In twelve or thirteen miles from where we crossed it, I sighted some pointed hills nearly to the eastward of us, and soon found, that the creek went between two of them, forming another pass. There was a waterhole under the rocks at this place also. This waterhole had no doubt been of large dimensions at one time, but it was gradually getting filled up with sand; there was however a good quantity of water there, and it was literally alive with fish, insomuch that the water had a most disagreeable and fishy taste, great numbers of dead fish floating on the water. I believe water can always be got here, as probably there is a permanent drainage from the creek into this hole. Here we found a considerable number of natives; and though the women and children would not come close, several of the men did so, and made themselves useful by holding some of the horses' bridles and getting some firewood. Most of them had names, given them by their godfathers at their baptism—that is to say, either by the officers or men of the O. T. construction at their first meeting. I called this place, which is my 32nd camp, "Rogers' Pass," after Murray Rogers, Esq., of the Darling-River, who is a partner of the Mr. Middleton, mentioned in yesterday's report, and who has so often treated me with the greatest kindness and hospitality. We made a good day's stage to-day, having travelled twenty-two miles.

Wednesday, 13th November.—From Rogers' Pass two conspicuous semi-conical hills—or, as I suppose I should say, two truncated cones—of almost identical appearance, caught my attention. They bore nearly S. 60° E., and the creek appeared to run close near them. After bidding adieu to our sable friends, who again made themselves useful by holding the bridles, while we were packing, &c., we started towards the two little conics, which I have named "The Twins"—and called one "Mount Reginald, the other "Mount Alfred"—after the twin sons of the late respected Major Campbell, formerly of Adelaide. To the south of The Twins—and of which they formed indeed a part—was a range of some length, having a red conic hill at its western point; I called this "Seymour's Range," after the well-known Mr. Robert Seymour of the Western District of Victoria, and also of South Australia. He has also had considerable experience in exploring portions of the north-west coast on the De-Grey, Fortescue and Ashburton Rivers. The red conic hill, at its western extremity, I named "Mount Ormerod," after a young gentleman who was a schoolfellow of Mr. Carmichael's, and whose acquaintance I made at Mr. Seymour's station. We passed the Twins in eleven miles, and continued down the creek, until we found some water about a mile from the eastern end of Seymour's Range, where there was a very peculiar red sandstone mount, which I named "Mount Quin," after Mr. Edward Quin of the Darling-River. The general bearing of our course to-day was nearly S. 70° E., and we travelled twenty-one miles in a straight line from the last camp.

Thursday, 14th November.—Soon after leaving the camp this morning, and passing Mount Quin, we remarked, that Seymour's Range, with Mounts Quin and Ormerod, had a series of lines of watermarks in horizontal rows all along them; the same was the case with Johnston's Range, which was the first range, which I had seen upon the Finke-River after starting from Chambers's Pillar. Seymour's and Johnston's Ranges lying almost E. and W. of one another; the latter range we were now again rapidly approaching. Soon after passing Mount Quin, but upon the opposite side of the creek, I found several fine ponds of water in a lateral clay channel; they were evidently permanent; I called them "Peddle's Ponds," after a kind and generous friend on the Darling-River, Mr. J. G. A. Peddle. The course of the creek from here was nearly eastward; and having taken my latitude this morning by a meridian altitude of Aldebaran, I was sure of what I before anticipated, that I was running down what I called my No. 2 creek, as on my outgoing march a creek had joined the Finke from the west near that camp. I travelled only fifteen miles to-day, and found a small fenced-in waterhole, which had been so fenced by the natives. There was a low range to the S.W., and a tent-shaped hill more to the eastward; a little to the west of north was a range I called "Bacon's-Range," after Mr. Harley Bacon, who had been in charge of all the stock during the construction of the overland telegraph, whom I met with Mr. McMinn at the Finke, and whom I also knew upon the Darling-River.

Friday, 15th November.—I rested the horses at this place to-day and did not move the camp. I walked to the top of the tent-hill, and from there saw, that the creek went through another pass a little to the N.E. of our camp. In the afternoon I rode over to this pass, and found some ponds of water a little to the west of it; a bullock, whose tracks I had seen on the creek, had got bogged here, and was now left high and dry. I called these ponds and pass "Briscoe's-Pass" and "Briscoe's-Ponds," and the little tent hill I have named "Briscoe's-Tent," after Mr. H. H. Briscoe, of the Darling-River, who was living with my two friends, Messrs. Middleton and Rogers, when I last saw him.

Saturday, 16th November.—This morning we again made a start from the fenced-in waterhole; we went through Briscoe's-Pass, which was only five miles from the camp. The course of the creek from this camp was considerably to the north of east, being indeed nearly north-east. In six miles from the pass the creek ran by a hill, which I well remembered in going out, and at eight miles its existence ended in the broader bosom of the Finke-River; there was a fine waterhole at the junction, in the bed of a smaller creek, but too boggy for horses to water at. The river appeared very different to when we passed up it; it then had a stream of water running down its channel; it was now dry, except that small patches of water showed above the sandy bed at frequent intervals, which were, generally-speaking, salty and bitter. Some few were drinkable enough. Upon reaching the river I camped. My expedition was over. I had failed in my object (to penetrate to the source of the Murchison-River) certainly, but not through any fault of mine, as I think any impartial reader of my journal will admit.

Sunday, 17th November.—This morning, at starting from camp (which had been about one mile and a-half from my outgoing No. 2 camp) we crossed the river and struck into the line of our outgoing march. Our old horse-tracks were very indistinct, but yet recognisable. We reached my old No. 1 camp in thirteen miles.

Monday, 18th November.—To-day we travelled to the eastward along the course of the River Finke, and passed a few miles to the south of Chambers's Pillar, which had been my original starting-point. I had left it but twelve weeks and four days to the time I resighted it, and during that interval I had traversed and laid down about 1000 miles of previously totally unknown country. My expedition thus early ended, but it was solely in consequence of having no one to accompany me. Had I been fortunate enough to have fallen upon a good or even fair line of country, the distance I actually travelled would have taken me across the continent. On the 21st November we reached the telegraph line at the junction of the Rivers Finke and Hugh, and here was the last I saw of my two late companions; they preceded me down the country. The weather during this month, and almost to its close, was infinitely cooler than the preceding one.

ERNEST GILES.

[On reference to the accompanying map it will be observed, that one of the culminating points in the MacDonnell-Range, discovered during this expedition, was named Mount Sonder, in honour of Dr. Will. Otto Sonder of Hamburg, who largely elucidated the marine *algæ* of Australia, and who also much contributed towards the knowledge of the land plants, particularly of the south-western portion of our continent. Two other high mountains in the same range were dedicated to Count Zeil and Baron von Heuglin, who distinguished themselves by recent geographic explorations in Spitzbergen, the latter also by his extensive zoologic researches in North Africa.—F. v. M.]

II.

MR. ERNEST GILES'S

GEOGRAPHIC EXPLORATIONS,

1873-4.

Monday, 4th August, 1873.—The party for Western Australia under my leadership, consisting of four members, viz., myself, Mr. William Henry Tietkens, Alfred Gibson, and James Andrews, with twenty-four horses, four riding and twenty packed, left the Main Overland Telegraph-Road at Ross's Waterhole, at the junction of the Stevenson and Alberga-Creeks, and proceeded up the latter to where the Telegraph-Line crosses it, being distant from Ross's Waterhole thirteen miles, and lying nearly west. The small frame of an old building was convenient for turning it into a house, with a tarpaulin for a roof; and as there appeared a likelihood of more rain (a few showers having fallen on the first instant) it was brought into service; water was obtained in some clay-pans in the immediate neighbourhood, and there we encamped for the night.

Tuesday, 5th August.—This morning was very cloudy, and some light mists warned us to have all our things under canvas. I was loth indeed to remain inactive so soon after effecting a start; but as some of the horses were missing, and rain had begun to fall, we roofed in the old building, and there we remained for the rest of the day; rain falling at intervals throughout, the clay-pans became quite full, and the sandhills soft with the moisture. At sundown heavy thunders and lightnings came from the north-west, and we had one closing shower of considerable volume; the remainder of the night was dry.

Wednesday, 6th August.—The morning rose fine and clear, though the night had been extremely cold. We effected a reasonably early start. I intended to follow up this creek, believing it would prove a well-watered one; its bed was broad but ill-defined, and cut up into numerous small channels. A low piece of table-land bore south 78° west, with apparently good sandhill-country round it; and as I expected that the creek would run nearly in that direction, I travelled to it about five miles. Thence travelling nearly west we entered a more scrubby region; and though the soil was soft from the rains, no clay-pans could be seen, nor was there any other surface water lying about. We struck the creek again in ten miles; but so little clay existed near it, that we had some trouble in obtaining any water at all; at length, however, we found a small clay-pan with only just sufficient. The creek seemed smaller and less likely to hold water than lower down; as not sufficient rain had fallen, to flood it, I concluded that it was but a poor watercourse after all, and as the country now appeared scrubby on both sides I determined to strike away more northerly or north-westerly for some of the creeks discovered by Stuart.

Thursday, 7th August.—The night was cold and dewy; the thermometer standing at 37°. A start through the straying of the horses was not effected until ten o'clock; then leaving the Alberga we steered away north-west. The country was covered with thick mulga-scrubs, and consisted of grassy sandhills, from the tops of none of which could any view be obtained. At four miles we reached a clay-pan with water in it, and gave our horses a drink, as there had not been sufficient at the camp; what little had been there over-night had been absorbed before the morning. We travelled twenty-four miles on this course through the same kind of country, but met with no more clay-pans; so at night had to camp on the top of a sandhill without water even for ourselves—though we were not destined to leave this spot without any, for at ten o'clock p.m. a thunderstorm came up again from north-west, and before we could get half our things under canvas we were thoroughly drenched, and thus we obtained sufficient water for breakfast; but the ground was too sandy to retain any.

Friday, 8th August.—This morning the horses were in several mobs in the scrub, and it was again late before we effected a start. Continuing on the same course, that is to say north-west, and through the same kind of country, we sighted some table lands at five miles, and at sixteen more we came down out of the sandhills on to a creek (the Hamilton of Stuart), where after a short search we found water in its bed. We camped here, but the feed was exceedingly poor, dry and innutritious.

Saturday, 9th August.—Last evening the sky again threatened rain, but the night passed without any, and the morning broke bright, clear and beautiful. Two Hamiltonians visited the camp this morning, and did not appear at all inclined to leave it. The creek here is broad and sandy bedded; the timber is small and stunted. The camp was not moved to-day. Towards evening the two Hamiltonians put on an air of impudence, and became rather objectionable, and two or three times I had to resent their encroachments into the camp, until at last they greatly annoyed me, and as our guns wanted cleaning and discharging, after the recent showers, as soon as they saw us handle and heard us discharge one, they departed.

Sunday, 10th August.—Although the day was Sunday (as we had rested the previous day), we departed from the little waterhole and proceeded along the north-bank of the creek. Two timbered ridges approached the creek on a bearing of N. 65° W. We passed the foot of them in seven miles, crossing a small tributary as soon as we left the camp and another at the ridges; the last-mentioned one had several fine clay-holes, all full of water, and no doubt it remains there for several months after rains; there were also many stony clay-pans full of water, which from the summit of the hill, I could see glittering in the sunshine. Another hill lay west, and the main creek evidently ran under its foot. On reaching it we crossed another tributary, with plenty of water in it, and found a large hole in the main creek, but it was perfectly dry; the natives had, when the water was there, fenced it in to catch the large game which might come to drink; but at present they were saved the trouble, as game and water had both departed. Mr. Tietkens, my lieutenant and second in command, had found some surface-water in a very pretty little amphitheatre, formed by the hills, and we encamped at it, having travelled 13 miles; the grass was wretchedly dry and dead, though the horses seemed to get on pretty well with it; so I suppose it was better than it looked. On the south side of the creek the country consists of red sandhills with scrubs behind them.

Monday, 11th August.—Starting at 7.30 this morning we continued up the creek and round the foot of the northerly hill which forms a part of the amphitheatre. From its summit, which was not very high, I noticed that a junction of another creek occurred there, and concluded it to be the place where Stuart crossed it; the creek which joined it came from the north, while the main-creek came more directly from the west, indeed its bend was a few degrees south of that point. We continued up the main channel; the bed was here broad, sandy and flat, with no signs of water anywhere on the surface, the timber growing all over the channel; the country on both sides consisted of mulga-scrub, more or less thick. We crossed and recrossed the bed, when having found a place where the natives had dug for water, though at only 12 miles from last camp, I determined to try, if water could really be got by digging, as, if not, it would be useless to follow this creek any further. We obtained the water easily enough, but the supply was not satisfactory; and an enormous quantity of sand had to be shifted, before we could water even the most willing of the horses. Most of them would not descend into the pit, which they doubtless supposed to be a trap set to catch them. The trouble and annoyance of horses, when to be watered at such places, cannot be either imagined or described; it must be felt, and then it is never forgotten. We succeeded at length, with the aid of canvas buckets, to water them, and by the time the whole twenty-four were satisfied, we were also; the grass was dry as usual, but the horses ate it with apparent satisfaction. Our course to-day was 8° south of west. Close to where we encamped, Mr. Tietkens found three or four saplings placed in a row in the bed of the creek and a small tent-frame, as though someone—if not done by native children—had been playing at erecting a miniature telegraph-line. I thought it might, perhaps, have been a freak of that man *Hume*, who so easily blinded the Sydney authorities as to let him out of gaol; as I had heard that he occasionally leaves the telegraph-stations and returns with some wonderful accounts of his travels and golden discoveries.

Tuesday, 12th August.—In consequence of the recent showers having left rainwater lying on any ground that would hold it (that is to say, clayey or stony), and not liking this creek any more than I did the Alberga, I decided to leave this also and try the country still further N.W. So this morning, leaving the creek and driving immediately into mulga-scrub, we continued for 18 miles. At 11 miles a number of the so-called poplar-trees were seen. At the end of 18 miles however, we came full upon the bank of the creek again, much to my surprise. I noticed also with regret a few (the first outposts) bushes of porcupine-grass (the

Triodia irritans). As usual there was no water in the creek. We followed it up five miles further, during which distance its course was north 20° west. It getting late in the day, we commenced digging for water, which was obtained at 3½ feet, at which depth we struck white pipeclay. The water supply was again very limited, and only our riding-horses could be watered by night, so little coming in, that a large hole was requisite. The day was exceedingly warm, in fact quite oppressive. From a low rise in the scrub I noticed the horizon was broken by some higher ridges, then bearing N. 30° W., and I supposed, if the creek trended up to them, the bed might become stony and surface-water be obtained; but in such a dense bed of sand, as I was now in, it could not be expected, as it could not remain a moment without sinking away.

Wednesday, 13th August.—Last night was cloudy and oppressive, and I had difficulties in obtaining observations, which however resulted in giving me latitude 26° 27′, longitude 134°. The camp was not moved, as it took nearly the whole of the day to water the horses. The day was again warm, the thermometer rose to 92° in the shade at three p.m. There was a warm breeze blowing from the north all day. The only incident which occurred out of the common to-day, was the fall of a long dead limb of a tree, which we had erected to stretch a tarpaulin, to give us a little shade, and which during dinner was blown down. It fell first on the head of Jimmy Andrews, the youngest of my party; also across my back, tearing my waistcoat, shirt and skin; but as it only fell on Jimmy's head, of course it couldn't hurt him.

Thursday, 14th August.—Last evening, at eight p.m., the thermometer had not fallen below 80°, and the night was close and oppressive; by daylight it had only descended to 67°. The country on either side of the creek being so scrubby, and the horses parting company, and having to water them only one at a time, when they were got, prevented an early start, but getting under weigh at length we travelled still up the creek on the same course as yesterday, namely, north 20° west: in a few miles we reached a low stony rise in the scrub, and from it I obtained a view of the creek, stretching away towards some other ridges, nearly on the same line we were going; so we continued skirting the creek, and reached the hills in eleven miles. From the same low hill I noticed, that nothing but scrub could be seen to the north or south, and only to the west-north-west was the horizon broken by hills, and they were of greater elevation than I should have expected. Reaching the ridges we found a place, where water could be easily got at a few inches in depth; so we encamped, as we might go farther and fare worse. The entire depth of the creek-bed here was not more than a foot, there being pipeclay at the bottom. The weather to-day was less oppressive, than the two previous ones. From the ridges, under which we encamped, I obtained a more extended view than formerly, and found the horizon from north-west by west round to north bounded by hills apparently of a greater altitude, than any I had yet seen; there were many long flat-topped hills and several singular cones. I was much pleased to have at length distanced the scrub, of which we had had so many miles, as a change of country appeared to occur here.

Friday, 15th August.—Last night was again clear, cold and dewy, though the thermometer at eight p.m. stood at 74°; by daylight it had declined to 56°. From the ridges above the camp one cone in the new range bore north 52° west, and for some distance the creek trended that way. Steering away therefore in that direction, we crossed the creek once or twice, and at length left it on our left hand. On reaching the foot of the ridges, we found the creek had greatly altered its appearance, if indeed it was the same creek; it now only appeared like a blind water-flat, with an odd eucalyptus-tree here and there, but with marks of heavy floods in its channel. It is possible, that the main-creek may have turned more west, and disappeared behind some of the scrubby ridges, but nothing could be seen from any of them; so we had to content ourselves with what we had; and as we found some surface water in a small clay-hole, we found it far more convenient than digging for water even in a larger creek. We travelled only eleven miles to-day, but here for the first time for many weeks we came upon some green grass, and the horses greedily devoured it; so we camped. The country for the last three or four miles had been very good, the scrub opening out, and the grass was much fresher. There were a few eucalypts near where we encamped, and the channel of the creek seemed much more defined a mile or two up; so it might yet turn out a respectable watercourse.

Saturday, 16th August.—Last night at dusk the thermometer stood at 65°, the night being clear and cool, but not dewy. At daylight it had fallen to 50°. On mustering the horses this morning one of them was found to be dead-lame with a stake in his fetlock, and it was quite impossible for him to travel; so we were forced to remain at this camp, though Mr. Tietkens and I mounted our horses and rode further up the creek. The channel recovered its appearance, and we concluded it was still Hamilton's-Creek. Its course above our camp was nearly north, and a line of bluff ridges formed its eastern bank. The country opened out towards the west, that is to say: in the direction of the new ranges, first seen on Thursday. There was no water in the creek as far as we went, but we found clay-pan water in two or three places. We next rode to a prominent little cone in the new range five or six miles from the creek. To the west-north-west the country was excellent, being open, grassy, and having fine cotton and saltbush flats all over it, and we saw plenty of clay-pan water. I called this the Anthony-Range, after His Excellency, the newly-arrived Governor, this being the first geographical feature, found upon this expedition. The day was delightfully cool and pleasant. We returned to the camp much pleased with our ride.

Sunday, 17th August.—At 8 p.m. last night the thermometer stood at 54°; the night was clear, dewy and extremely cold. At daylight the thermometer indicated 34°. The lame horse was very bad again this morning; but we lightened his load as much as possible, and he travelled pretty well after the first mile. Leaving the creek at last for good on our right hand, we struck through a small patch of scrub, and entered an open valley, which led up into the hills of the Anthony-Range. I steered for the singular cone, seen some days ago; it was now only about ten miles off. Most of the hills on this range are flat-topped, and many tent-shaped hills exist also. Our course was a little west of north-west. On reaching the cone, which I ascended, I obtained a view of the surrounding country, which displayed hills for several miles in all directions, amongst which also were many bare rocks of red colour, heaped into the most fantastically tossed mounds imaginable, with here and there an odd shrub growing from out the interstices of the rocks. Some small miniature creeks, with only myall and mulga growing on them, ran through the valleys, all of which had recently been running; we therefore had no difficulty in finding sufficient surface water, and we encamped a mile or so beyond the cone, in an extremely pretty and romantic valley. The grass was green, and Nature appeared in one of her smiling moods, throwing a sunshine on the minds of the adventurers, who had sought her in the wilderness. The only miserable creature in our party was the lame horse. I said he was the only one; but we soon found that another horse had staked himself during our day's march, though he did not go lame until his hobbles were put on, when we found we had two patients instead of one; Mr. Tietkens extracted a long mulga-stick from his fetlock. (Neither of these horses ever became sound again, though they did their work always well enough.) The distance travelled to-day was twelve miles.

Monday, 18th August.—Last evening about dark the thermometer stood at 50°, and there were signs of a very cold night approaching, which proved to be the case, for by morning the mercury had descended 2° below freezing point, and though there had been a heavy dew, there was neither frost nor ice. I was glad this morning to find that the worst of the two lame horses had joined the mob in the night, and was feeding contentedly with all four feet upon the ground. We got under weigh this morning at an early hour, and passed up to the head of the picturesque valley, in which we had encamped, and from there we turned round some of the mounds of bare rocks, previously mentioned; they are composed of a red kind of conglomerate granite. We wound in and around hills, until at length we reached the bank of a small creek, lined with eucalypts, and finding some water in clay-pans near the bank, we encamped on a piece of beautiful country, being splendidly grassed, and ornamented with the fantastic mounds and creek-timber as back and foreground for the picture. We had travelled only thirteen miles, but the country was so agreeable, that we had no great desire to traverse it at great speed. From the last hills, climbed by us, another and independent range, about 35 miles away, bore west-north-westward, and I was grieved to notice, that the intermediate country seemed all scrub.

Tuesday, 19th August.—Last night the thermometer stood at 58°; this morning at 34°. The night was clear and cold, but not dewy. Making an early start, we crossed the little creek, which had no water in its bed whatever; the grass along its banks was beautifully green. Travelling in a straight direction for the distant range, discovered yesterday, we soon entered a thick mulga scrub, and on it I noticed a plot or two of the dreaded triodia and a clump or two of the small mallee. At ten miles we came upon the bank of a very large creek, with fine and vigorous eucalyptus timber. Mr. Tietkens soon found a native well with water in it at no great depth. The course of this creek, where we struck it, was to the south-south-east. I intended to camp here, but Mr. Tietkens persuaded me to follow it up a bit, as he remarked, that such a large creek, going in anything like our direction, should not be lost sight of; it coming from the north-north-west induced me to travel upon that course, but that line took us immediately into thick scrub, and away from the creek altogether, so at four miles, not seeing anything of it, I was compelled to climb a tree, and saw its timber some distance away, and noticed that I must turn north to cut it, which I did, and in three miles we came upon another creek at right angles, which I felt sure could not be the one we had left, but must form a junction near here, but the scrub being so thick, we could hardly see a yard a-head. Here I sent Jimmy Andrews up a tree; having been a sailor he is well skilled in that kind of performance, but I am not. I told him to discover the whereabouts of the main creek, and say how far it appeared. That brilliant youth informed me that it lay across the course which we had been steering, and was only a mile away, so of course we went on to it (as we thought), but having gone more than two miles, and not reaching it, I asked Jimmy if he had not made some mistake, as I said "We have already come two miles north, and you said it was scarcely a mile;" he then kindly informed me, that I was going all wrong, that I ought not to go that way at all; but upon questioning him as to which way I should go, he replied "Oh, I don't know *now;*" so my only resource was to turn east, and then of course we struck it in a few miles, then Jimmy declared, if we had *kept north long enough* we would have come to it *again*. Where we now struck it the channel was by no means the size we expected to see it, which rather puzzled me, for I could not believe that the mere addition of the branch-creek, crossed in the scrub, could make so large a channel as we had originally discovered. We were compelled to conclude, that we were still only upon a branch, and that the large main creek was still farther east; however it was too late to go much farther, and we tried to get water by digging, but were quite unsuccessful. We had travelled nearly twenty miles, and had to camp without any water. I made up my mind in the morning to return to where we had originally struck the large creek, which I named the Krichauff, after the Honourable F. H. Krichauff, of Adelaide. As we had seen water at about three feet depth in the native well there, I now was sorry that I had not camped at it, when I first intended. The latitude of this camp was 25° 56′.

Wednesday, 20th August.—The thermometer last evening about dusk stood at 46°, intimating a cold night, it being 12° colder than the previous night. In the morning the mercury was down to 28°, but no ice or frost appeared. This morning Mr. Tietkens, when out after the horses, found a native well some little distance down the creek with water in it, though it was deeper than we were usually accustomed to; so we packed up and shifted the camp down to the native well. On the way going, on the outer bank of the creek, I happened to be behind the party, and before I overtook them I heard the report of firearms. On reaching the party, Jimmy Andrews had his revolver in his hand, but Mr. Tietkens was away; and upon inquiring of Jimmy the cause of the reports, and the reason of his resorting to his revolver, he replied, that he thought Mr. Tietkens was shooting blacks, so he got his pistol ready. Mr. Tietkens had been firing at some wallabies. On reaching the well we had a vast amount of work to perform, so that only three or four horses got water by night. I was not only sorry but angry with myself for not having returned to the large creek, where what was required could have been obtained so much easier. Mr. Tietkens indeed worked most enthusiastically at this tank, returning to it and working by himself after all the rest had retired. I was sorry to see him expend his strength in almost useless labour; for I told him that I should not wait a moment in the morning, but decamp to where I knew water could be got with comparatively speaking little labour, as the horses (except four) had now been two nights without water; but Mr. T. having worked so hard and so long during the night, he was actually able to water all the horses in the morning. The weather to-day was most delightful, but the night set in much warmer than last.

Thursday, 21st August.—At dusk last night the thermometer stood at 56°; by daylight it had reduced its indication by one-half. There was a heavy dew fall, but no frost or ice; and in consequence of the dewy moisture on green feed the horses were not so anxious for water as might have been expected, though when it came to actual drinking one would think that they had been a week without. I was delighted to turn my back upon this wretched creek and tank, and we travelled away on a west course, crossing the western tributary in two and a-half miles. The object of our present course was to reach the new hills, seen from the Anthony-Range. Three of them appeared higher and isolated from the lower ones. They now bore west of us—at least, they should have done so; and I hoped they did; for in such thick scrub it was impossible to see them. No matter for that, we steered west for them, and traversed a region of dense scrub. I was compelled to ride in advance with a bell on my horse's neck, to enable the others to know which way to come. We had started rather late, and striking a small dry creek with eucalypts in the scrub at seventeen miles we encamped, there being reasonably good herbage here. In the scrub to-day we came past a native pheasant's or lowan's *(Leipoa ocellata)* nest, but there were no eggs in it.

Friday, 22nd August.—The night was again cold, clear and dewy. At dusk the thermometer stood at 46°; at daylight at 26°. There was a slight hoar-frost this morning. Making an early start, we proceeded still west, still through scrub, but a little less thick than yesterday. At six miles we reached the foot of some low broken ridges, the last two or three miles having been open and beautifully green. From the top of this ridge we had before us a most charming piece of country, red ridges of the most extraordinary shape and appearance tossed up in all directions, with the slopes of the sandy soil from whence they seemed to spring rising gently, and covered with verdure like a green carpet up to their feet. These slopes were most beautifully clothed with grass, herbs and flowers of the most varied hues, throwing a magic charm over the entire scene. Vast bare rocks—

> " Piled on rocks stupendous hurled,
> Like fragments of an earlier world,"

—appeared everywhere; but the main tier of ranges, for which I had been steering, was several miles further to the west. But thinking, that water—the scarcest of Nature's gifts—must surely exist in such a lovely region as this, it was more with the keen and critical eye of the explorer in search of that element, than of the "admirer of Nature in her wildest grace" that I surveyed the scene. A small eucalyptus creek lay to the south, to which Mr. Tietkens went. I sent Gibson to a spot about two miles off to the west, as straight before us in that direction lay a huge mass of stones and slabs of rock, which might have rocky waterholes amongst them. To the north lay a longer jumble of hills with overhanging ledges and bare precipices, and I went to search amongst them, leaving Jimmy to mind the horses, until some of us returned. Though

Mr. Tietkens took the shovel, he could get no water by digging. Gibson also returned unsuccessful. I had travelled for miles in and out, and over hills and rocks, through gullies and under ledges, still without success, and was returning quite disappointed, when I chanced to come upon a small and very fertile little glen, whose brighter green attracted my notice. A small gully only came down between two hills, and meandered through it. I only expected a rock-hole, but greatly to my surprise, in the bed of the little channel, and in the midst of sandy ground, I saw a patch of black soil, and on reaching it found it to be a small native well with a little water at the bottom. It was an extraordinary spot, and being funnel-shaped, I doubted if any animal could get down to it. I also expected it would be a perfect bog; but my horse appeared so determined to test its nature, that I let go his bridle, and though it was over three feet deep, he quietly let his forefeet down into it, and helped himself to a good drink, though his hind quarters were high and dry above his head. I had great doubts if he could get himself out; but the soil being reasonably firm, when he had satisfied himself, he wriggled out somehow. I immediately returned to our rendezvous, where the others had returned some time. I led them up to where I had found the water, where we camped, having travelled only nine or ten miles. We had of course immediately to set to work to enlarge the well, and we found the water supply was by no means abundant; for though we all worked hard in turns at the shovel, only half the horses could get water by night, as it did not drain in as fast as one horse could drink; but we expected, by making a large hole, sufficient would drain in during the night to water the remainder of the horses. We did not cease our work until it was dusk, when we retired to our encampment, quite sufficiently tired to make us sleep without the aid of any lullaby.

Saturday, 23rd August.—The night set in very cold, the thermometer about dusk standing at 46°, but the night was not nearly so cold as the two previous ones, as the thermometer had not fallen below 40° by daylight. Upon inspection this morning we found that but a poor body of water had drained into the well during the night, and that there was by no means sufficient for the remaining half of the horses which had none yesterday. We passed the forenoon in still enlarging the tank, so that a bucketful could be given to some of the horses as it drained in. Mr. Tietkens and I took our horses and rode off to the main hills to the west, in the hopes of discovering more water. Before we left, however, Mr. T. and Gibson had planted in the good soil near the tank a quantity of seeds, such as of blue-gum, wattle, melons, pumpkins, cucumbers, maize, &c., being some from a collection, sent me by my kind friend Baron von Mueller. Being late when we started, it was dusk when we reached the main range. The country passed over between it and our encampment was exceedingly beautiful, hills being thrown up in red ridges of bare rock, with the native fig-tree growing amidst the rocks, festooning them into infinite groups of beauty, whilst the ground upon which we rode was a perfect carpet of verdure. We were, therefore, in high anticipation of finding some waters equivalent to the scene. But as night was advancing our search had to be delayed until the morrow. The dew was falling fast, the night air was cool and deliciously laden with the scented exhalations from shrubs and flowers. The odour of almonds was most intense, reminding me of the perfumes of the wattle-blooms of the eastern and more fertile portions of this continent. So exquisite was the perfume, that it recalled to my mind Gordon's beautiful lines—

> "In the spring when the wattle gold trembles
> 'Twixt shadow and shine,
> When each dew-laden air draught resembles
> A long draught of wine."

So delightful, indeed, was the evening that it was late when we gave ourselves up to the oblivion of slumber.

Sunday, 24th August.—At 8 p.m. last night, the thermometer had registered 46°; but had not descended below 40° by daylight. We had made a fire against a large log, about 18 inches through, being a limb from an adjacent blood-wood tree, and this morning we discovered that it had actually been chopped off its parent stem with either an axe or tomahawk, and that it had been carried forty or fifty yards away from where it had been cut. By sunrise we were upon the top of the mountain; it consisted of enormous blocks and boulders of red stone, so riven and fissured, that no water could lodge for an instant upon it. I found it also to be highly magnetic, and it turned the needle from its true north point to 10° south of west; but the attraction ceased when the compass was removed four feet from contact with the rocks. The view from this mount was of singular and almost awful beauty. The mount, and all the others connected with it, rose simply like islands out of a vast ocean of scrub. The beauty of this locality lay within itself. Innumerable red ridges, ornamented with fig-trees, rising out of green and grassy slopes, met the eye everywhere to the east, north and north-east, and the country between each was sufficiently timbered, to add charm to the view; but signs of water did not increase the cheerfulness of the landscape, nor any basin or hollow, which could hold it, met the gaze in any direction; that alone was wanting to turn a wilderness into a garden. There were four large mounts in this chain, higher than any of the rest, including the one I was on. Far away to the north some ranges appeared above the dense ocean of intervening scrubs. To the south scrubs reigned supreme; but to the west, the region for which I was bound, the horizon looked far more cheering. The far horizon there was bounded by a very long and apparently connected chain of considerable elevation, seventy to eighty miles away; one conspicuous mountain, evidently nearer than the longer chain, bore 15° to the south of west; an apparent gap or notch in the more distant chain bore 23° to the south of west. The intervening country appeared all flat, and very much more open than that in any other direction. I could discern long vistas of green grass dotted with yellow flowers; but as the perspective declined, these all became lost in lightly timbered country. My mind was immediately made up; I would return at once to the camp, where the water was so scarce, and trust all to the newly discovered chain to the west. Water must surely exist there; we had only to reach it. We descended from this mountain, having first named it Mount Sir Henry, after the Hon. Sir Henry Ayers, K.C.M.G. The other three large mounts in this range, lying to the northward of this, I named respectively Mount Barrow, after the Hon. John Barrow; Mount Reynolds, after the Hon. Thomas Reynolds; and Mount Cavenagh, after the Hon. William Cavenagh. The whole range and all the hills connected with it, I have named Ayers-Range. On returning to our camp, Gibson was away after the horses. Only a mere dribble of water was in the tank. On Gibson's return with the horses, he informed us that he had found another place with some water lying on the rocks and close by two native wells, much shallower than our present one, and each with water at the bottom, and that they were three miles away. I rode off with him to inspect this new discovery, and found there was sufficient surface water on the summit of one of the rock mounds for all the horses for a day or two. These rocks indeed are most singular, being mostly large red round solid blocks of stone, shaped like the backs of enormous red turtles. I was much pleased with Gibson's discovery, and we returned to move the whole camp down to this spot, which we always after called the Turtle-Back. As far as grass and herbage were concerned, the horses could not have been in better country. It is wonderful, that so little water exists here; but the horses having been curtailed in their water supply had not benefitted so much, as they would otherwise have done. The surface water here was rather difficult for horses to reach, as it was upon the extreme summit of the Turtle-Back, and the sides were very steep and slippery; but after the first attempt they would come and climb up of their own accord. Having the two native wells to fall back upon, I thought it a good opportunity of visiting some of the nearest western mountains, while the horses had a few day's rest. Mr. Tietkens planted a quantity of seeds at this place, in the rich soil round the native wells. Round upon the opposite or eastern side of the Turtle-Back was a large ledge or cave, under which the natives frequently encamped. It was ornamented with many of their rude representations of creeping things, among which the serpent class predominated; there were also other hideous shapes of things, such as can exist only in their imaginations, and they are but the weak endeavours of these benighted beings, to give form and semblance to the symbolisms of the dread superstitions, that haunting the vacant chambers of their darkened minds, pass amongst them in the place of either philosophy or religion.

Monday, 25th August.—This morning, after seeing all the horses watered and having a fine bath on the top of the Turtle-Back, Mr. Tietkens and I got our nags and started for Mount Henry, lying nearly west. Reaching it, we travelled upon the bearing of the gap in the most distant range, which we had formerly seen. The country as we proceeded we found agreeably open, splendidly grassed, and it rose at intervals into a few low stony ridges. At fifteen miles from the Turtle-Back we found an open piece of plain, with some clay-pans to our right; and on searching about we saw a small supply of extremely thick fluid, all ploughed up with emu tracks. We turned our horses out here for an hour, during which time several emus came round us. Mr. Tietkens had a pop at one with his revolver, and hit it in a fleshy part of the neck, which did not prevent it running away at full speed. Re-saddling we continued our journey, proceeding first up the plain for half-a-mile; then we had some continuous well-timbered ridges before us. They ran mostly in parallel lines, north and south, having splendid little grassy valleys between them, down which water courses, in times of rain, to the south. There were no creeks or eucalyptus-timber; but a few well-grown beefwood-trees *(Grevilleas)* ornamented the scene. At twenty miles from the Turtle-Back we left these behind. The country now opened out into beautiful grassy land on the south, while to the north and infringing upon our track, more low well-grassed ridges ran along. We passed a small watercourse, with here and there a well-grown eucalyptus-tree; but the channel was perfectly dry. Proceeding on, as evening was beginning to close in, we observed the eucalypts of another creek, whose timber indeed would scarcely reach our track; so, as it would not come to us, we went to it. The timber on it was thick and vigorous, and it excited hopes of the discovery of water. The creek came from a north-westerly direction. We found the so-called teatree (*Melaleuca*) growing here. Following it up for about two miles we reached a place, where a low ridge existed on both sides of the creek, and here we found an old native well, which, upon cleaning it out with a quart pot, disclosed the element of our ardent search to our view, at a depth however of nearly five feet. Here we encamped, having travelled twenty-five miles. The natives always make these wells of such an abominable shape—always that of a funnel—never thinking how awkward to white men with horses. It took us some time to water our horses, Mr. Tietkens doing that part of the work, while I collected firewood and made some tea. The horses (always thirsty when there is any trouble in watering them) drank nearly four buckets each, when they departed to graze in as fine pasture as ever they saw. There was a quantity of the little purple pea-vetch, of which they are so fond, and which is so fattening for them. There was plenty of this also at the Turtle-Back, and wherever it grows it gives the country a lovely carnation-tinge, which blending with the bright green of the young grass and the yellow hue of several kinds of flowers, impresses on the whole region the appearance of a garden.

Tuesday, 26th August.—The night was cold and dewy. The horses declined to drink this morning. Passing over a low ridge, so as to regain our course, we continued on for ten miles, when we noticed, that the nearest mountain, seen from Mount Sir Henry, was now not more than thirty miles away. It appeared red, bald and of some altitude. To our left and south-eastward was another mass of jumbled turtle-backs. Where we got the water last night we should have had to dig out a large hole, to water all our horses, and we thought to find some surface water in the turtle-backs to our left, with the view of saving us the trouble of digging; so as they were not far away we went to search amongst them. On ascending one we noticed a small gum-creek away to the south-south-east; and, after failing to find water amongst these hills, we went away to the creek, which was equally destitute of water. We had now wasted so much of the day unsuccessfully, that we determined to leave none of these hills unvisited; but having searched until late, we could only return to the little well, left by us in the morning. I was greatly annoyed with the loss of our day. None of the hills had the slightest sign of a rock-basin; indeed Nature seemed to have made no effort to form any such thing, and we had seen no place whatever, where the natives had ever dug. We had been riding from morning till night, and had neither reached the first mountain nor found water. We had to dig out the native well again with the quart-pot, and our horses at length went off to feed on the richest vegetation. I determined in the morning, to send Mr. Tietkens back to the Turtle-Back and bring the party on as far as this creek, while I went to the mountain, to see how the country looked there. We now discovered that we had brought but a small supply of provisions with us, and that a heavy supper would make a desperate hole in it, so it was necessary to take a supper that was easily digested, which we accordingly did.

Wednesday, 27th August.—The night was again clear, cold and dewy. Getting our horses early and watering them, we departed each on his separate errand—I for the mountain, Mr. T. for the camp. I made a straight course for it, and in three or four miles found the country became exceedingly scrubby. At ten miles I came to a number of native huts; they were of large dimensions and two-storied. It was strange to find them in the scrub; but then, as the natives can get water out of the roots of the mulga, it was not so surprising after all. As I proceeded the scrub became thicker and thicker, and I could not see the mountain, except at intervals. At twenty-two miles I reached a spot where the natives had burnt the old grass, and new and rich vegetation was now growing there, so I determined to give my horse the benefit of an hour's aest. The day was exceedingly fine, the thermometer standing at 76° in the shade. Having had but a poor breakfast, I had an excellent appetite for all the dinner I could command, and could not help thinking, that there is a great deal of sound philosophy in the Chinese doctrine, "that the seat of the mind and intellect is situate in the stomach." Starting again, and gaining a rise in the dense ocean of scrub, I got a sight at the mountain, whose appearance was most wonderful; it seemed so rifted and riven, and having acres of bare red rock without a shrub or a tree upon it. In five miles from where I had rested my horse, I suddenly found myself under the shadow of a huge rock, towering above me amidst the scrub, but which had been so completely hidden, as not to be seen until I reached it. On ascending it, I was much gratified to discover, not more than a mile and a-half from me, the eucalypts of a creek, which meandered through this dense wilderness. I soon gained its banks, but was disappointed to find, that its channel was very flat and but poorly defined, though the timber upon it was splendid—elegant upright stems, supporting umbrageous tops—whose roots at least must surely extend downwards to a moistened soil. On each side of the creek was a strip of open ground, so richly grassed and so beautifully decked with flowers, that it seemed like suddenly escaping from purgatory into paradise, when emerging from the recesses of the dense scrub on to the banks of this beautiful creek. Upon the side opposite to where I struck it, stood an extraordinary ridge or hill, consisting of a huge red turtle-back, having a number of enormous red stones, almost egg-shaped, traversing (or rather standing in rows upon) the whole length of the ridge. I could compare it to nothing else than an enormous oolitic monster of the turtle-kind, carrying its eggs upon its back (not that there was any limestone in the neighbourhood). There were a few cypress-pines growing among the interstices of the rock, which gave it a most elegant appearance. Hoping to find a rock water-hole, I rode towards this hill, but before reaching its foot I came upon a small piece of open, firm, grassy ground, most beautifully variegated with many-coloured vegetation, with a small bare piece of ground in the centre, with rain-water lying on it. The place was so exquisitely lovely, it seemed as if only rustic garden seats were wanting to prove, that it had been laid out by the hand of man. But it was only an instance of one of

Nature's freaks, and in which she had most successfully imitated her imitator, Art. I watered my horse and left him to graze on this delightful spot, while I climbed the turtle's back. There was not much water here—not sufficient for all my horses; so it was actually necessary for me to find more. The view from this hill was wild and strange. The high bald forehead of the mountain was still four or five miles away, and the country between was all scrub. The creek came from the southwestward, and was lost in the scrub to the east of north. There was a thick and vigorous clump of eucalypts down the creek a mile or two; so I decided first to visit them, and accordingly rode down to them; but the bed of the creek was perfectly dry. Retracing my steps, I then searched up the channel, and tried by scratching in many places; but no water rewarded my labours. Then, continuing still further up, I came to a place, where great boulders of stone crossed the bed, and where several large-sized holes existed, but they were now perfectly dry. Continuing my search near these, to my great delight I at last found a damp spot, and near it a native well, not more than two feet deep, and having water in it. Further up again I found an overhanging rock with a good pool of water underneath it, and the sand all round quite wet; and water appeared by just kicking the sand away. I was perfectly satisfied with my day's work, as a good supply of water could be got here without any trouble. Here I decided to camp. As there was a large log lying in the creek, I could make a good fire at it. The horse was up to his eyes in most magnificent herbage, and I quite envied him his supper. Natives had been here very recently, and the scrubs were burning not far off to the north. As night descended, I lay down by my bright fire in peace to sleep; though doubtless there are many of my readers, who would scarcely like to do the same. Such a situation might naturally lead one's thoughts to consider, how many people have lain similarly down at night in fancied security, to be awakened only by the enemy's tomahawk crashing through their skulls. Such thoughts, however, did not intrude themselves upon my mind. My thoughts wandered away to different scenes and distant friends and relations; for this

"Childe Harold, also, had a mother not forgot,
And sisters whom he loved, but saw them not,
Ere yet his weary pilgrimage begun."

Dreams, also, between sleeping and waking, came rushing through my brain—

"Dreams of that which cannot die—
Bright visions—came to me,
As lapped in thought alone I'd lie
And gaze into the midnight sky,
Where the sailing clouds went by
Like ships upon the sea.

"Dreams that the soul of youth engage
Ere its passions have been quelled—
Old legends of the monkish page,
Traditions of the saint and sage,
Tales that have the rime of age,
And chronicles of eld."

Then in my lonely sleep I had real dreams—sweet, fanciful and pleasant dreams, mostly connected with the enterprise I had commenced—dreams that I had wandered into and was passing through tracts of fabulously valuable plains, watered by never-failing supplies of the purest of crystal, dotted with clusters of magnificent palm-trees, and having groves—charming groves—of the fairest of pines. On awaking, however, I was forced to reflect, how

> "Mysterious are these laws:
> The vision's finer than the view:
> Her landscape Nature never draws
> So fair as fancy drew."

Thursday, 28th August.—Last night the thermometer stood at 56°; this morning it had declined to 28°; and now

> "The morn was up again—the dewy morn—
> With breath all incense, and with cheek all bloom,
> Laughing the clouds away with playful scorn,
> And smiling as if earth contained no tomb,
> And glowing into day——"

With this charming extract from Byron for breakfast, I saddled my horse, having nothing further to detain me here, intending to bring up the whole party as soon as possible. This time I returned by a more southerly route, and found the scrub less thick. At thirteen miles I came to some low red rises, hid in the scrub, having travelled nearly east. From here, I made a direct line for the Teatree-Creek, expecting to meet the whole party there. At six miles on the new course I came upon an open piece of ground, and saw several emus, which induced me to look about for water; I soon found some clay-pans with sufficient to last for a week. I was much pleased with my discovery, as it would be a capital place to camp at with the whole party *en route* to the new creek, and it would save us the trouble and accompanying bother of digging out a tank at the Teatree-Creek. The water here was certainly rather thick, and scarcely fit for human organisms—at least for white ones, though it might suit black ones well enough; it, however, was good enough for horses, and that was the great consideration. I rested my horse here for an hour, and again started for the Teatree. On reaching it the party had not arrived, and I had to wait in expectation of their arrival. In about an hour Mr. Tietkens came and informed me, that on his road back the other day he had found a nice little waterhole about six miles from here, and that the party were camped at it; so we returned in company to it. It appeared that the water supply at the main camp at the old Turtle-Back had been quite exhausted, when Mr. Tietkens brought the party away; I was therefore all the more pleased in having found more in advance.

Friday, 29th August.—The night was again clear, cold and dewy, the thermometer at night standing at 48°; in the morning at daylight it was down to 30°, but there was no frost. We made an early start from this very agreeable little spot, where there was plenty of wood, water and grass, and travelled to the thick water clay-pans, which I had stumbled across yesterday, without troubling the little Teatree-Creek any more, reaching the clay-pans in nineteen miles.

Saturday, 30th August.—Last night the temperature of the atmosphere was warmer than usual, the thermometer at 8 p.m. not having fallen below 60°; at daylight it was no lower than 40°, and the sky was cloudy. We got away early from camp, and reached my former solitary encampment at the rockhole early in the afternoon. We got plenty of water for the horses with very little trouble, and they were soon amidst splendid vegetation. This was an auspicious day in our travels, being no less than Mr. Tietkens' nine-and-twentieth birthday. We celebrated it with what honours the Expedition stores would afford, obtaining a flat bottle of spirits from the medical department, with which we drank to his health and many happier returns of the day. I also, in honour of the occasion, called this Tietkens' Birthday Creek, and hereby proclaim it unto the nations, that such should be its name for ever.

Sunday, 31st August.—The thermometer last night stood at 56°, this morning at 28°. The camp was not moved to-day. Mr. T. and I however rode over to the high mountain, taking with us all the apparatus necessary for so great an ascent—that is to say, barometer, thermometer, compass, field-glasses, quart-pot and water-bag.' We reached its foot in about four miles, and found its sides so bare and steep that I took off my boots for the ascent. It was formed for the most part like a stupendous turtle-back, of a conglomerate granite, with no signs of any water, or any places, that would retain it for a moment, round or near its base. Upon reaching its summit the view was most extensive in every direction, except the west; and though the horizon was bounded in all directions by ranges, yet scrubs filled the entire spaces between. To the north lay a long and very distant range, which I took to be the Gill's-Range of my former expedition. A little to the west of it was another range, but both were so indistinct even with the glasses that I could not conclude, whether they were near ridges or distant ranges. To the north-westward was a flat-topped hill, rising like a table from an ocean of scrub. It was very much higher, than such hills usually are. I have since learned that it was called Mount Conner by Mr. Gosse. To the south and at a considerable distance lay another range of considerable length, apparently also of some altitude. This I have named the Everard-Range, after the Hon. Wm. Everard. The western horizon was bounded by a continuous mass of hills and mountains, stretching from N. 73° W. to S. 15° W., and from the centre of which Birthday-Creek seemed to issue. Many of the mounts in the range to the west appeared of considerable elevation. The natives were burning the scrubs to the north and north-west. On the bare rocks of this mountain I picked up several of what I at first thought were sea-shells, like those of the periwinkle, but I soon discovered they were merely old and empty snail-shells. We had first seen them at Mount Sir Henry. I was grieved to find, that my only barometer had met with an accident in climbing this mountain, and that it was now perfectly useless; however by testing the boiling of water I hoped to get the altitude of this mountain. I found the water boiled at 206°, showing the elevation to be 3790 feet above the level of the sea. The boiling point of water at the camp on Birthday-Creek was 208° 30′, showing the elevation of the mountain above the surrounding country to be 1295 feet. The view of Birthday-Creek winding along in little bends through the scrubs down from its parent mountains, was most pleasing. Down below us also were some very pretty little scenes. One was a small sandy channel, like a plough-furrow, with a few eucalyptus-trees upon it, running from a ravine near the foot of this mount, which at three-quarters of a mile passed between two red mounds of solid rock, only just wide enough to admit its course, with a few cypress-pines growing close to the little gorge. On any other part of the earth's surface, if such another place could be found, water must certainly exist also, but here there was none. We had a perfect bird's-eye view of the spot. We could only hope, for beauty and natural harmony's sake, that water must exist at least below the surface, if not above. Having completed our survey, we descended barefooted as before, and decided to follow up the creek we were now camped on, though it is not probable that its course will be much more than twenty or thirty miles. On returning to the camp we found Gibson and Jimmy had shot some parrots, and we had an excellent dish of parrot-soup. We have up to the present seen only a few small magpies and one or two pigeons, though plenty of kangaroos, wallabys and emus, but have not been successful in getting any; they are exceedingly shy and difficult to approach. I named this very singular mount the Sentinel, as I soon found:

"The mountain there did stand
To sentinel enchanted land."

Monday, 1st September.—Last night the thermometer fell to 40°, and by morning it stood at 26°. There was very little dew, but what there was became slightly frosted. The camp was not moved to-day, as I had a great quantity of work on hand. I found my position here to be lat. 26° 3′, long. 132° 39′ E.

Tuesday, 2nd September.—The thermometer last night was up as high as 60°, and by daylight had not fallen below 56°. Heavy clouds were flying past last evening, and during the night a few drops of rain also fell at intervals. At about 10 o'clock I observed in the northern horizon, what I do not remember ever having seen before, namely, a lunar rainbow. Its diameter was only about 15°; there were no prismatic colours visible about it. To-day was clear and fine, but rather windy. We travelled up the creek, skirting its banks, but cutting off the bends. There were some low ridges on our right. The course of the creek for some distance was S.W., then more southerly; but at ten miles it made more directly for the hills to the west. The country along its valley was excellent, and the scenery most beautiful, pine-clad red and rocky hills being scattered about in various directions; while further to the west and south-west the high, bold and very rugged chain rose into peaks and points. We travelled sixteen miles, and encamped close to a pretty little pine-decked hill on the north bank of the creek, where some rocks traversed the bed, and we easily obtained a good supply of water at a very shallow depth; and as the grass and herbage were magnificent, the horses were in a fine way to enjoy themselves. Really, this spot is one of the most charming the imagination could paint. In the back-ground were the high and pointed peaks of the main-chain, from which sloped a delightful green valley, through which the creek meandered, here and there winding round the foot of little pine-clad hills of unvarying red colour, whilst the earth from which they sprung, was covered with a carpet of verdure and vegetation of almost every imaginable hue. It was happiness to lie at ease upon such a carpet and gaze upon such a scene; and it was happiness, most ecstatic, to know I was the first white man who had ever beheld it. My visions of a former night really seemed to be prophetic.

Wednesday, 3rd September.—The thermometer last night stood at 50°; by morning it was down to 40°. A very cold wind set in this morning, which made the atmosphere feel as though the thermometer had fallen far below freezing-point. The trend of the creek and valley, down which it came, was nearly 25° south of west. Travelling up it, we found it became contracted by impinging hills. At ten miles we found a small pool of water in the creek-bed. From here in about a couple of miles to my surprise I found we had reached its source, which was the drainage of a big hill. There were no more water and no rock-holes, neither was there any gorge. The hills were clothed with triodia, but none grew on the lower ground. The valley turned to a charming amphitheatre. Thus we had traced our Birthday-Creek to its own birthplace. It has but a short course, but a merry one, and had ended for us at its proper commencement. There appeared no likelihood of any other water at hand, though there were plenty of clayholes and channels in the amphitheatre, which were all dry; so I elected to return to the pool we had previously seen about two and a-half miles back, where we encamped. Several small branch-creeks running through beautiful little valleys joined our creek to-day, but all were dry. We were now near some of the higher cones of the main-chain, and could see that they were all entirely timberless and covered with triodia, which however grew only upon the hills. The spot we were now encamped upon was another scene of exquisite sylvan beauty.

Thursday, 4th September.—Thermometer last night 60°, this morning 46°; the night being windy and cold; dense nimbus-clouds hovering just above the mountain-peaks threatened a heavy downpour of rain, but the driving gale scattered them into the gelid regions of space, and after sunrise we had a perfectly clear sky. At starting this morning I intended to push through what seemed now (as it had also seemed from the first moment I saw this range) a main-gap through the chain. Going north first round a pointed hill we were soon in the trend of the pass. In five miles we reached the banks of an eucalyptus-creek coming from the hills on our left or south of us, and running down towards the main-valley further west. We thought it might perhaps be worth a tank or two, so we did not altogether despise it, though it was small. In another mile, however, we saw that it ran either into another creek or else a large eucalyptus-flat, which had no apparent outlet. This heavy timber could be seen for two or three miles. Advancing still futher I soon discovered, that we were upon the reedy banks of a fast-flowing stream, whose murmuring waters, ever rushing idly and unheeded on, were now for the first time disclosed to the delighted eyes of their discoverer. Here I had found a spot where Nature truly had

> "Shed o'er the scene her purest of crystal, her brightest of green."

This was really a delightful discovery. Everything was of the best kind here—timber, water, grass and mountains. In all my wanderings, over thousands of miles in Australia, I never saw a more delightful and fanciful region than this, and one indeed where a white man might live and be happy—my dreams were of a verity realised. Geographically speaking, we had suddenly come almost upon the extreme head of a very large watercourse. Its course was nearly south, and I found it now ran through a long glen in that direction. We saw several fine pools and ponds, where the reeds opened in the channel, and we flushed up several flocks of ducks. It appeared to me, that the water was at a rather low ebb; for, although there had been rain enough to start young grass and new herbage, yet not nearly sufficient to flood any creeks had fallen, for not one of the branch-creeks, which ran into Birthday-Creek, had the slightest appearance of having been flooded for years, nor was there a drop in the large holes at our first camp on that creek, which if filled would probably hold for a year; therefore I think I saw this region by no means at its best—that was the opinion I formed upon the spot; at all events I thought it high time that such a locality should be no longer nameless, I therefore christened it upon the spot—"The Ferdinand Chain," after my best and kindest friend, Baron von Mueller, a principal subscriber to the fund for my expedition, and who furnished also many of the instruments. I have since discovered, that it was seen from a distance and named by Mr. Gosse the Musgrave Ranges. The watercourse, which commences its existence here and ends perhaps far away, I have named "The Ferdinand;" and the glen where I now am, and through which it leaves its parent hills, I have named "Glen Ferdinand." The glen extended nearly five miles; and, where it ended, the water ceased to show upon the surface. At the end of the water we encamped, and I do not remember any day's work during my life, which gave me more pleasure than this; for I trust it will be believed that—

"The proud desire of sowing broad the germs of lasting worth
Shall challenge give to scornful laugh of careless sons of earth.
Though mirth deride the pilgrim-feet that tread the desert plain,
The thought that cheers me onward is—I have not lived in vain."

We had travelled only about ten miles, but that was quite far enough in so excellent a country as this. After dinner Mr. Tietkens and I ascended a high mountain to the S.S.E. of the camp; not that it was the highest in the neighbourhood, for several others not far away were higher, but it was the most convenient; water boiled at its summit at 204°, while at the camp in the glen below it was a trifle over 206°. I call this "Mount Ferdinand," and another higher point nearly west of it I called "Mount James Winter," in honour of James Winter, Esq., of Rushworth, Victoria, one of the subscribers to the fund for my expedition. The view all round from west to north was shut out. To the south and south-west other ranges existed. The timber of the creek could be traced for many miles in a southerly direction. It finally became lost by distance in a timbered country, not to say scrub. This mountain was highly magnetic. Very little native signs were visible in the glen. We returned to the camp and sowed the seeds of many cereals, fodder plants and vegetables. A great quantity of tea-tree grew in this glen; the water was pure and fresh.

Friday, 5th September.—The thermometer last night stood at 50°, this morning at 36°. At starting this morning we went down the creek two or three miles to where it passed between two red hills. I was now compelled, by the configuration of the mountains, to take a south-westerly valley for my road, which however sooned turned more to the W.N.W. At seven miles we crossed the channel of a fine timbered creek, which came out of the high hills at the foot of Mount James Winter, and which joined the larger one in a few miles. Where we crossed it the bed was dry, but the sand damp. On riding up it about a mile I found plenty of running water, but it did not run quite so far down as our crossing-place. It was a fine creek; I called it the "Reid," after Samuel Reid, Esq., of Wangaratta, Victoria, one of the subscribers to the expedition-fund. We now found ourselves traversing another very pretty valley, running about S. 70° W., with fine cotton and saltbush flats upon it, picturesque pine-cypresses covered hills on both sides of us. At thirteen miles our course was obstructed by hills running across the western horizon. Under them we came to the banks of a small creek. There was no water just where we struck it, and it might have been some distance up it to where there was any. As we had only come a short distance, and the grass and herbage were so excellent, the horses did not require any; except that water was wanting, it was an exceedingly pretty place. The latitude of this camp was 26° 24'.

Saturday, 6th September.—The evening was very cool, and the thermometer at night stood at 44°; by morning it had fallen to 34°. The horses rambled a long way last night, and Gibson lost the only one he found near, or let it get away from him somehow; so the morning was entirely gone, before we left the camp. We proceeded down the creek for about two or three miles nearly south, then leaving it and passing through a little gap, we were enabled to travel west. At about three miles we arrived upon the banks of another fine creek, with good timber on it. The natives were burning grass or bushes a short distance up, and I was sure of plenty of water where they were. We therefore turned up the creek, which here ran north-westerly. Mr. Tietkens and I rode on in advance, to reconnoitre and look for a good camping-place. We had to go nearly three miles up it, when we came to water running along its bed, and at the same time we evidently disturbed a considerable number of natives, who were there, and who raised a most frightful outcry at our sudden and unexpected appearance amongst them; they did not run away, but walked deliberately into the bushes, which were very thick just here, and consisted of tall saltbush and teatree. While watering our horses a great many ran at us, poising their spears, which were 10 to 12 feet long, and of which each individual had an extraordinary number. When they saw us remain quietly sitting upon our horses, which however became very impatient, they began their horrible yells and gesticulations, some ordering us away, others climbing into the nearest trees and directing their spears at us from the branches; another lot, now on the opposite side of the creek, came also rushing up with spears advanced and "ensigns spread," and with yells encouraged those near us to spear us; but they had some doubts of the vulnerability of the horses. However, at the head of our new assailants was one man sophisticated enough to be able to call out to us—"*Walk, whitefellow—walk!*" but as we remained immobile, he induced two others to join him in making a rush at us, and they hurled their three spears at us before we could get out of the way. It was fortunate we were at the extreme distance from them that these weapons can be projected; they fell or struck the ground right at our horses' hoofs, which made them very restive. I thought it was time to let them see, that we were not quite as helpless as we appeared. I immediately unslipt my Snider-rifle, and not aiming at them, or intending to kill or wound any of them, I fired at random, and the bullet went so scientifically between two of these three worthies and smashed some boughs just behind them, that it instantly produced silence amongst the whole congregation, but only for a moment. All this time we were most anxiously awaiting the arrival of Gibson and Jimmy with the horses; but these valiant retainers preferred to remain down the creek, where we had left them and out of harm's way, leaving us to kill or be killed, as the chances of war might determine. We at length had to retreat from our sable enemies and go and find our friends. We finally got the whole lot up and watered the horses, but the yelling of these fiends in human form, the clouds of smoke from the burning grass and bushes, and the many disagreeable odours incident to a large native village, all combined to make both us and our horses exceedingly restless. The natives seemed quite overawed by the number of horses; and though they crowded round in all directions (for there were more than 200 of them) they did not then throw any more spears. I selected a small open piece of ground, back from the water, to camp upon, though the hills came down rather closer to it, than I could have wished. When they saw us dismount and commence unpacking, they commenced properly to work themselves up for a regular onslaught. As long as the horses remained close they seemed disinclined to attack; but when the horses were all hobbled and went away, they made a grand sortie—rushing down the hills (just at the back of the camp and where they had all congregated) towards us in a body, with spears in pose and yelling their war-cry. Of course by this time, seeing our lives were in real danger, I ordered out all the firearms we could muster, which amounted to two Sniders, two shot guns, and four revolvers, so we were quite prepared, when they charged us. I ordered that for the first volley the two Sniders and one gun should be discharged, and to fire at the ground near their feet. I delayed almost a second too late, for at the instant I gave the word to fire, five spears had left the enemy's hand; but our shots cutting up the ground at their feet and sending sand and dust all over them made the great body of them pause. We then ran at them and fired our revolvers in quick succession, and what with the noise, and the to them extraordinary phenomenon of a projectile approaching them, which they couldn't see, we drove them up into the hills again, and they were quiet for nearly an hour, during which time we improved the occasion by getting some dinner. That meal however was not completed, when we saw them stealing down on us again, for as we had not hurt any of them, they came to the conclusion I suppose that we could not. Again they made a rush, with heads back and arms at fullest tension, to give their spears proper projective force, when just as they came within spear-shot—for we knew the exact distance now—we gave them another volley, striking the sand up just before their feet; they again halted, consulted one another by looks and signs, when the discharge of Gibson's gun with cartridges in it decided them, and they ran back; but only to come again. In consequence of our not wounding any of them, they began to jeer and laugh at us, slapping their backsides at us, and jumping about in front of us, daring and deriding us and our weapons. We were compelled at last to send some Snider-bullets into such close proximity to some of their limbs, that at last they really did believe we were dangerous folk after all. Toward night their attentions ceased, and though they camped just opposite to us, they did not trouble us any more. Of course we kept a watch during the night. The men of this nation were big, exceedingly hirsute, and in good bodily condition; there were none with grey hair amongst them, and their features were by no means prepossessing; some wore the chignon, others not; some were painted with red ochre. I did not notice any boomerangs amongst them. I named this encounter-creek the Officer, after C. M. Officer, Esq., of Victoria, one of the subscribers to the expedition-fund. There was a high mount to the north-east from here, and lying nearly west from Mount James Winter, which I have named Mount Officer, after S. H. Officer, Esq., of Victoria, another subscriber; the sources of this creek, I imagine, come from it.

Sunday, 7th September.—Though there was a sound of revelry by night in the enemy's camp, ours was not passed in music. The evening had been cold, the thermometer being down to 44°. By morning it stood at 34°. We were not much troubled with the natives this morning, but as some horses were in sight, Gihson got one to ride after the others with, and Mr. Tietkens took Jimmy with him on top of a hill, to take some bearings for me, and immediately they found me alone, they recommenced their malpractices. Having my arsenal in good fighting order, I determined, if they persistdd in attacking me, to let some of them know, what a white man could do if he liked, but as only two or three came close, I fired in the air. They fell back, and though they kept close round watching us, and yelling perpetually, they did not attempt any further attack. I was very gratified to think afterwards, that no blood had been shed, and that we had got rid of our enemies with only the loss of a little ammunition. Leaving Gibson and Jimmy to pack the horses, Mr. T. and I rode up the creek to see, what water supply there was here—the creek had a fine open valley, and there were plenty of reeds in patches in the creek-bed. The water disappeared at a mile above the camp, and for a mile the bed, though damp, was dry—at least it had no surface-water on it. We then came to where it was running again for more than a mile; and so on alternate spots of dry and running continued as far as we went. At all events there is water in plenty for all the stock, which this country could possibly carry, and it is undoubtedly permanent. When we returned to the camp, the horses were packed, and though it was Sunday, I determined not to remain here any longer, as the horses were frightened at the incessant yells of the natives, and our ears, at least mine, were perfectly deafened with their outcries. We departed at 1 p.m., leaving the aboriginal owners of this splendid piece of land in the peaceful possession of their beautiful hunting-grounds. We went up a little valley nearly west, and through a small gap at three miles, passing a low timbered hill on our left at seven miles, and having the main-range immediately to our right, or north of us. It continued stretching away in heavy high masses of hills, and with a fine open country to the south. At ten miles we crossed another fine creek with excellent timber. Its bed was dry, where we crossed it. While the party went on, I rode up it and came to running water in about a mile and a half, and no doubt for miles beyond that the bed is full of water. I named this fine creek the Currie, after John Currie, Esq., of St. Kilda, one of the subscribers to the expedition-fund. It was rather late, when we arrived at the next creek, being six miles further on. Turning up it a bit we soon found water, and encamped. This was another fine watercourse, with young and vigorous timber. I called it the Levinger, after Bernhard Levinger, Esq., of Melbourne, another subscriber. The country passed over to-day was excellent, being fine open grassy valleys all the way.

Monday, 8th September.—The night was warm, the thermometer standing at 60° at night, and at 40° in the morning. The horses being close handy, we got an early start. Keeping a short distance away from the foot of the main-range, which of course lay to the north of us, we continued for twelve miles, the course being about W.N.W. The country to-day was rather more timbered, than we had previously met with in this range. At twelve miles we reached the banks of another fine water-course; the horses being almost unmanageable with flashness, running about with their mouths full of rich herbage, kicking up their heels, and biting at one another, that it was almost laughable to see them, with such heavy packs on, attempting any such elephantine gambols; so I determined to give them a good day's stage, to steady them a bit. I went by myself up this creek a short distance, and came to the water running down from the mountains and disappearing in the sand; further up towards the hills no doubt there were miles of running water. I called this Winter-Water, and the creek in which it exists Murndale-Creek, after S. P. Winter, Esq., of Murndale, Victoria, a subscriber to my fund. The creek had a most inviting valley up in the hills; but I did not ascend it further, than where I met the water. At seventeen miles from our camp at the Levinger, and about five from the Murndale, we passed a very high mountain in the range, which seemed to me to be the highest I had seen in it. I named it Mount Davenport, after the Hon. Samuel Davenport, of Adelaide. We now passed up through a small gap over a low hill, and immediately on our appearance we heard the yells and outcries of more natives down on a small flat below, where we saw a small and I suppose happy family, consisting of two men, one woman and one youth, but whether of the male or female sex, I could not determine. When they saw us descend from the little hill, they very quietly like respectable people walked away. Continuing our course in nearly the same direction, that is to say, about W.N.W., and passing two little dry creeks, I climbed a small hill and saw a most beautiful valley about one mile away, stretching out to the north-west, with eucalyptus-timber up at the head of it. The valley appeared entirely enclosed by hills, and was a most enticing sight. Passing through two or three hundred yards of mulga, we came out on the open ground, which was really a sight, that would delight the eyes of a traveller even in the Province of Cashmere, or any other region of the earth. The ground was covered with a thick carpet of green grass and herbage; conspicuous amongst the latter was an abundance of the little purple vetch, which, spreading over hundreds of acres of ground, gave a lovely pink or magenta tinge to the whole scene. I found also, that there was another valley running nearly north, with another creek meandering through it, and apparently joining the one, formerly seen in this valley, though the timber ceased for a mile or two, and grew again below the junction to the south. We had travelled twenty-three miles. I intended searching in the north-west creek for water, but in passing across this fairy space I noticed the white semi-sodaish appearances, that usually accompany springs and flood-marks in this region. I soon reached a most splendid kind of stone-trough, under a little stony creek, which formed an excellent spring running into and filling the little trough, running out at the lower end and disappearing beneath the surface, evidently perfectly satisfied with the duties it had to perform. This was really the most delightful little spot I ever saw;

a region like a garden, with springs of the purest water spouting out of the ground, for ever flowing into a charming little basin, which was one hundred yards long by twenty feet wide, and in the deepest part four feet deep. There was a quantity of the teatree-bush growing along the various channels, which all contained running water. The valley is surrounded by picturesque hills, and I am certain, that this is the most charming and romantic spot I ever shall behold. I immediately christened it the Fairies' Glen, for it had all the characteristics, to my mind, of fairyland. Of course we encamped here. I would not have missed finding such a spot upon—I won't say what consideration. There were numbers of both ancient and modern native huts near this water, and this is no doubt an old-established and favourite camping-ground of theirs. And how could it he otherwise? No creatures of the human race could view these scenes with apathy or dislike, nor would any sentient beings part with such a patrimony at any price but that of their blood. But the Great Designer of the Universe, in the long past periods of creation, permitted a fiat to be recorded, that the beings, whom it was His pleasure in the first place to plant amidst these lovely scenes, must eventually be swept from the face of the earth by others more intellectual, more dearly beloved and gifted than they. Progressive improvement seems the order of creation; and we, perhaps, in our turn, may be as ruthlessly driven from the earth by another race of yet unknown beings of an order infinitely higher—infinitely more beloved than we. On me, perchance, the eternal obloquy of the execution of God's doom may rest, for being the first to lead the way, with prying eye and trespassing foot, into regions so fair and remote; but being guiltless, alike in act or intention, to shed the blood of any human creature, I must accept it without a sigh.

Tuesday, 9th September.—The thermometer last night fell considerably lower than the night previously; registering 44° at night and 24° in the morning, at which time there was a quantity of ice on the water left in the billies and pannikins. I did not move the camp to-day; the place was so charming, I could not tear myself away. Mr. Tietkens and I walked up to and climbed a high mount about three miles north-easterly from the camp; it was of some elevation. We ascended by a gorge, having eucalypts and callitris-pines half-way up the mountain; we found water running from one little basin to another, and high up, near the summit, a bare rock, over which water was gushing. To us, as we climbed towards it, it appeared like a monstrous diamond flashing back the rays of the morning sun. I called this Mount Oberon, after Shakespeare's king of the fairies. The view from the summit was limited. To the west, the hills of this chain still ran on; to the east I could see Mount Ferdinand. The valley, in which the camp and water were situate, lay in all its enchanting loveliness at our feet, and the little natural trough in its centre, now reduced in size by distance, looked like a silver thread, or indeed it appeared more as though Titania, the queen of the fairies, had for a moment laid her magic silver wand upon the grass. The day was lovely, the sky serene and clear, and a gentle zephyr-like breeze merely agitated the atmosphere. As we sat with a lingering gaze over this delightful scene, and having seen also so many lovely spots in this chain of mountains, I was tempted to believe I had discovered regions, which might eventually support not only flocks and herds, but centres of population, each individual among whom would envy me as being the first discoverer of the almost paradisiacal scenes, it so delighted them to view. In the afternoon we returned to the camp, and again and again wondered at the singular manner, in which the water existed here. Five hundred yards above or below it there is no sign of water; but in the intermediate space a stream gushes out of the ground, fills a splendid little trough, and gushes into the ground again—emblematic indeed of the ephemeral existence of humanity; we rise out of the dust, flash for a brief second in the light of life, and in another we are gone. We planted seeds at this spring. I called it Titania's Spring, and the creek in which it exists Moffat's Creek, after Robert Moffat, Esq., of Ravenswood, Victoria, a contributor to the fund for my expedition.

Wednesday, 10th September.—The night was much warmer than the last, the thermometer not having descended below 66° at night and 56° by morning. The horses, upon being brought up to the camp this morning, displayed such abominable liveliness and flashness, that there was no catching most of them. Fortunately I just managed to secure one colt, who was the leader of the riot; the others had to be driven several times at a gallop round the camp before their exuberance had in a measure subsided. It seemed indeed, as if the fairies had been bewitching them during the night. It was consequently late when we left the Fairies' Glen; a pretty valley, running north-west, with a creek in it, was our road. Our track wound about through the most splendidly grassed valleys, mostly having a trend westerly, when at twelve miles we saw the gum-timber of a watercourse, apparently debouching through a glen. Of course there was water there, and a channel filled with reeds, down which the current ran in never-failing streams. The spot was another of those charming gems, which exist in this chain, and I called it Glen Watson, after J. B. Watson, Esq., of Sandhurst, Victoria, another contributor to the fund for my enterprise. From a hill near this camp I discovered, that this chain had now become broken, and though it continues to run on still farther west, it seemed as though it would soon end. The Mount Olga of my former expedition now lay about 17° west of north from here, and appeared a considerable distance away. I had a great personal desire to visit it, as I had made many endeavours to do so on my former journey, but was prevented; now however I hoped no obstacle would occur, and therefore intend to travel towards it to-morrow. There was more than a mile of running water in this glen; the horses were up to their eyes in the most luxurious vegetation, and our encampment was again in a most romantic spot. Ah, why should regions so lovely be traversed so soon!

Thursday, 11th September.—The thermometer last night stood at 48°, this morning at 34°. A quantity of dew fell last night, being most probably caused by the moisture in the glen, and not by atmospheric causes, as we have previously had none for some time. Not expecting to get any water in the country between here and Mount Olga, though there were some low hills in that direction, I did not leave this camp until late, so that the horses came to drink of their own accord. We then left the pretty glen, with its purling stream and reedy bed, and entered very shortly upon an entirely different country, covered with porcupine-grass. We travelled north-west to some low ridges, which we reached in seventeen miles, and passed through some small belts of thick scrub, but of no great extent. There was no water at the ridges, nor any place where it could lodge;

but the horses having so lately watered, and plenty of beautiful pea-vetch growing here, they did quite well without. I noticed to-day, for the first time upon this expedition, some of the desert-casuarina, of which I saw quantities on my former expedition.

Friday, 12th September.—We made an early start this morning, and proceeded nearly on the same course to a round pine-clad hill. The intermediate country being low ridges and good valleys, we reached the hill in nine miles. From it Mount Olga bore north, it was still a great distance away, and I was anxious to find water before reaching it, if possible, as I of course could not tell, if any existed there. From this hill I could also see, that the Ferdinand-Chain, though broken and parted in masses, still continued along to the west, and rose again into high mounts and points; but strictly speaking the two lines were separate; I therefore called the western continuation the Bowen-Range, after His Excellency the Governor of Victoria, Sir George Bowen, G.C.M.G. (called the Mann-Ranges by Mr. Goss); near the foot of this round hill I was fortunate enough to discover a small flat piece of rock, which was hardly perceptible amongst the grass; on it I saw a few dead sticks, and an old native fireplace, which excited my curiosity, and on riding up to it, found to my astonishment under the dead sticks two splendid little rock-holes or basins in the solid rock, with ample water in them for the requirements of all my horses. I have oftentimes found water in strange places, and many times when I little expected it; but this was the most surprising little treasure of all. Of course we camped here, though we had travelled but nine miles; most of that distance was covered with triodia, and there were some bits of scrub to go through. I named these singular basins Fraser's Wells, after Mr. W. Fraser, late of the Barrier-Range, and now of Wagga Wagga, a kind friend of Mr. Tietkens. Mount Olga appeared fifty miles off.

Saturday, 13th September.—The thermometer last night did not fall below 60°, by morning it had descended to 48°; we experienced a feeling in the air of an approaching warm day; there was a good deal of scrub round the camp, and the feed was poor here, so the horses came late and consequently a late start was made. Our course was north, but a short distance east of our line some hills stood, which we deviated to, but no water existed about them. At nine miles we passed up through a stony gap, and came upon a small eucalyptus-creek on the north side, but it was dry. A great number of the fine casuarina-trees were seen to-day, but they were scattered thinly over the country, which was covered with triodia; the ground was pretty level, and we continued travelling until sundown, having travelled twenty-seven miles; there being neither water nor grass, our horses had but a miserable night.

Sunday, 14th September.—The thermometer stood at 60° last night, and 50° this morning. Making an early start, we arrived at the foot of Mount Olga in twenty-one miles. Mr. Tietkens here pointed out to me the track of a dray or waggon, also of horses and camels. I knew at once, that they must be Mr. Gosse's, but I was perfectly dumbfoundered at their appearance here; had the earth yawned at my feet, for ever separating me from this mountain, or had another of similar appearance risen suddenly before my eyes, I could not have been more astonished at the sight; for I knew Mr. Gosse had left the Telegraph-Line many hundreds of miles to the north of my starting-point; and as the line from Mount

Olga to the Murchison had been officially pointed out to me, and as I had understood, that Mr. Gosse was to penetrate the country north of the MacDonnell-Range, and being so excellently equipped to traverse any region, it was no wonder I was astonished at the appearance of those tracks. I was not only astonished—I had reason to be annoyed as well—because, as the Government-Expedition had come down to this mountain, and was now travelling in advance of me, on the only line of country that seemed traversable—that is to say—along the line of range now lying south and south-westward from here, it had probably more than a month's start of me. I was compelled to reflect, of what earthly use was it for me to continue in the same region with another explorer ahead. I had thoughts of returning immediately, and throwing up my expedition at once. If I have displayed an unseemly warmth of expression I am sorry; but I consider I should be quite excused for so doing; for, having looked upon Mount Olga, as it were, for my real starting-point for the west-coast, I could have reached it by a much easier line (at least a known one) and all explored by myself previously, and thus saved myself many hard hour's and day's work on the route I have now come; but being partly equipped by private, and partly by official support, I considered it my duty, or at all events wished to show my zeal, by traversing an entirely new country to it—and on reaching it, having travelled nearly 400 miles through a previously unknown tract, the instant I arrive at it, I find that the South Australian Government-Expedition had come there also! It was almost a death-blow to my expedition. It was therefore with no very gratified feelings, that I encamped at this mountain, which I had vainly hoped, having been the first to discover, I should also be the first to reach. The appearance of this mountain is marvellous in the extreme, and almost baffles description. Here I shall only remark, that it is formed of several vast and solid rounded blocks of bare red-coloured stone, of a kind of conglomerate, being composed of untold masses of round stones of all kinds, mixed as plums in a pudding, and set in vast and rounded shapes upon the ground. There was water running from the base of one of the hills, down a stony channel, and filling several rocky basins. The supply of water is not great, but I consider it permanent. The water disappeared in the sandy bed of the little creek, where the solid rocks ended. When Mr. Gosse was here, the water had evidently ran much farther down the creek channel, as where he had camped at a small waterhole, a mile nearly from where it now ended, a small bloodwood-tree was branded $\overset{G}{\underset{S}{o}}$. I had seen several quandong-and so-called poplar-trees on my route to-day.

Monday, 15th September.—I did not remove the camp to-day; indeed I had not decided upon my future movements at all. I made an attempt to climb a portion of this singular mountain, but the sides are too perpendicular. I saw a small pile of stones, erected probably by Mr. Gosse, at an elevation of several hundred feet. We collected a quantity of quandongs here, and dried them in the sun. There were some very beautiful black and gold butterflies, with very large wings. The day was warm, the thermometer indicating 95° in the shade.

Tuesday, 16th September.—The feed not being super-excellent here, I gave the horses another day's rest. I had intended, upon reaching Mount Olga, to have had some short respite, and all along our route everything, that was wanted to be done, was always put off until Mount Olga was reached; but Mr. Gosse's tracks here have upset all my ideas. I thought I was the monarch of all I surveyed, and the lord of the fowl and the brute; but lo! a greater than I is here. So I must depart to some remoter spot, where none shall dispute my sway. I first however determined, that I would overtake Mr. Gosse, and have a friendly intercourse; then we might agree to differ in our lines of march. There was another peculiar-looking mount lying nearly east from here, from whence Mr. Gosse's track had evidently come; but which way he had gone from here, simply consulting the dray-track, I could not discover in consequence of its many turnings. However it was evident enough, that he would naturally travel to the line of ranges, which stretched across the southern horizon, and that I should certainly pick up his tracks either to the right or left of me in a few days.

Wednesday, 17th September.—I moved the camp only one and a-half miles to-day, to a well-grassed flat. There were some pointed ridges to the west-south-westward, with the long extent of the Bowen-Range lying far beyond them, where I expected to meet with Mr. Gosse. The appearance of Mount Olga from this camp was most singular. I can only liken it to several enormous rotund or rather elliptical shapes of rouge-mange, which had been placed beside one another by extraordinary freaks or convulsions of nature. I found two other running brooks, one on the west and one on the north side, our first and Mr. Gosse's camps being on the south side; the latitude of this mount being 25° 20′, longitude 130° 57′.

Thursday, 18th September.—After watering the horses at the mountain we made a reasonably early start, traversing a country entirely consisting of sandy soil, rising mostly into sandhills, though patches of level ground occurred occasionally. There were casuarinas and triodia in plenty, being two kinds of vegetation, which appear thoroughly to enjoy one another's company. At about two miles we crossed Mr. Gosse's dray-track, apparently going due west, but I continued on to the hills lying to the west-south-west. We travelled twenty miles, and camped without water in the porcupine-grass. The day was rather oppressive, the thermometer going up to 95° in the shade.

Friday, 19th September.—Getting away pretty early from this camp, we struck the most eastern portion of the hills, for which we had been steering, in eleven miles. Turning round them to the west, I found a small supply of water up in a small gully, sufficient only for one-third of the horses; but I thought, that by opening it out with the shovel we might get more towards the morning. We therefore camped, and worked hard in removing rocks and stones, and digging out holes all the afternoon, there being only a very slight drainage here.

Saturday, 20th September.—At daylight there was not sufficient water for more than half the horses; so I rode away to look for more, while my companions distributed what there was amongst the animals. There was a sugarloaf-hill here, and near it I found a small creek coming from a rocky gorge, up which I found Mr. Gosse's dray-track again; also on a ledge in an almost unreachable place, sufficient water for about half my horses, if they could only get at it. There had been plenty of water, when Mr. Gosse was here; but since that time it had all dried up. I hastened back to the camp, and eventually brought all the horses round to this place. It was a dreadfully rough, rocky place. They arrived too late to get any water to-night. It was about five miles from the other camp.

Sunday, 21st September.—Though it was Sunday, the water being entirely exhausted, the horses getting only half a drink each, it was absolutely necessary to leave here. I intended from here to steer for a point on the Bowen-Range, which lay to the south-westward, and started as early as possible. Mr. Gosse's dray-track was left upon our left-hand or to the eastward of our line. The country consisted chiefly of sand-hills with casuarina and flats with triodia. At five miles we crossed a low ridge, and I spent some time in searching for water, but without success. Overtaking the party again, we encamped for the night without water in a dense casuarina- and triodia-country. We passed through several patches of mallee to-day, and gathered a quantity of quandongs again also. We saw also some curragong- and poplar-trees. A few miles only of country upon our day's march were very good.

Monday, 22nd September.—We did not get away from this camp until late, as some of the horses rambled, and a great deal of time was lost in recovering them. There were some ridges on our line a few miles nearer than the main-ranges, which we reached in thirteen miles. They were ornamented with cypress-pines; and I found a small gully, where I thought water might be got by digging. I therefore determined to camp and give the place a trial, as the horses greatly wanted water. Gibson and I went up the gully with the shovel and set to work; but at night, though we had dug a large tank, only a very small quantity of water drained into it, therefore the horses could not have any by night. I could only hope that a larger quantity would drain in by the morning. The last two days had been exceedingly warm and oppressive. I had hoped for some showers of rain at the equinox, but none had fallen. The country all round and amongst these ridges was very good indeed, and there was plenty of the purple vetch growing here.

Tuesday, 23rd September.—The thermometer last night stood at 72°, by morning at 58°. I was greatly pleased this morning to find, that our dug-out tank was quite full, and would most probably water all the horses; it however only watered seventeen—the remainder had to wait, but by 12 o'clock all had been satisfied. I determined to push on at once to the Bowen-Range, which now stretched across the whole southern horizon from south-east to south-west of us; it was about eleven miles away. From the hill near the camp a very enticing-looking glen between two high hills lay S.S.W.; we therefore departed for it. We traversed casuarina-sandhills, thickly covered with triodia, then some belts of mallee and finally some mulga-scrub; we then came upon open ground two or three miles from the foot of the range. We crossed the bed of a kind of a swamp or waterflat, being the end of a gum-creek, which issued from the range to our left; the ground was still soft, and showed signs of having been recently flooded. We kept straight on for the glen; as we approached it, eucalyptus-timber appeared to issue from the range, thereby proving that a watercourse existed there. We arrived at the mouth of the glen in twelve miles, and found water running in all three of the channels, into which the creek had split; the water had a very slightly brackish taste. Here we encamped. The day was very warm, the thermometer having stood at 94° in the shade.

Wednesday, 24th September.—The thermometer last night stood at 74°;

and by daylight at 58°. Heat and cold are evidently relative perceptions, for though the thermometer stood at 58°, I felt the atmosphere exceedingly cold this morning; but upon the appearance of the orb of day all coolness disappeared. After breakfast Mr. Tietkens and I took a walk up the creek into the glen down which it flows, and up on to some hills, which environ it. We found the water running rapidly down the glen, and discovered several fine rock basins, where water might be obtained for a long period after the flowing had ceased, if indeed it ever does cease. At three-quarters of a mile up the glen we came upon one very fine reservoir, nine or ten feet deep; the pellucid element descending into it from a small cascade in the rocks above. It was the largest sheet of water *per se* I had yet discovered upon this expedition, and I believe it is always there. It was the most picturesque and delightful bath, and as we plunged into its transparent depth, we revelled in the extasy of an almost newly discovered element. I called this charming little spot Zoe's Glen, after the fair daughter of His Excellency Sir George Bowen, whose fame had reached even unto the confines of the wilderness. I must here remark, that when I named this the Bowen-Range, I had no idea of Mr. Gosse's advance of me, and that I really had no right to name it. I am therefore compelled to apologise to His Excellency, for having placed his name to a geographical feature without having the power to maintain it; however the name of the glen will for ever remain unaltered, though the range is called (as I have learned since my return) by Mr. Gosse, the Mann-Range, after the Hon. Charles Mann, of South Australia, I presume. The camp was not removed to-day, as I looked upon this as my Sunday. We did not read divine service, as I considered the whole of my pilgrimage a divine service. And in our wanderings up the glen, we had

"Books in the running brooks, and sermons in stones."

The latitude of this pretty little retreat was 25° 59′. The day was warm enough; the thermometer standing at 92° in the shade.

Thursday, 25th September.—The thermometer last night stood at 68°; this morning at 52°. We remained at this most agreeable little spot again to-day. There was a creek a mile or two to the eastward of this glen, and I rode over to inspect it. I found it to be a fine little creek, larger than that at the camp, with plenty of running water along its channel, and I travelled a long distance up it. I could discover no large waterholes, so that if the stream ceases to flow, water would not remain long there, though I believe water is always to be found here. I called this creek, Christy Bagot's Creek, after Mr. Christopher Bagot, of the Peake-Station. I flushed up several flocks of ducks, but having no gun could get none. Upon my return to the camp, having reported the ducks, Gibson and Jimmy walked over on a shooting excursion; but returned at night having only bagged three, which we turned into an excellent stew.

Friday, 26th September.—The thermometer last night stood at 74°, by morning at 62°.; there was a strong gale of warm wind blowing from the south all night. The horses being now quite refreshed, having been on excellent country for the last two nights, we departed again from Zoe's Glen and travelled first north-west for a mile, in which direction the hills of this range trended; we then travelled for three miles on a bearing of 26° north of west; then three miles 2° north of west, and here from a hill I observed a kind of valley between the ranges, running nearly south. I now turned down it; there were casuarinas and triodia in this valley near the north end, but it improved to the south—it formed a regular pass between two masses of the range. At 4½ miles down it Mr. Tietkens and I mounted a rocky rise, and he being in advance of me informed me, that there was a lake below us, two or three miles away. I was soon able to see it also, and was much gratified at the sight. It lay south 37° west from where we saw it. We immediately proceeded on towards it. The valley or pass had now become somewhat choked up with low pine-clad stony hills, and in less than a mile we came upon a running creek with some fine little sheets of water in it, and it meandered round these low hills and descended into the lake at last. The water was very slightly brackish. I am convinced these are permanent springs. I have called them the Hector-Springs, and the pass through the range Hector's Pass, after Hector Wilson, Esq., of Horsham, Victoria, a contributor to the fund for the expedition. On reaching the banks of the lake, I found its waters also slightly brackish. There was no timber on its banks—it lay close under the foot of the mountains, having their rocky slopes for its northern bank; and the bed, for probably half the distance over, is stony. The opposite shore was sandy; numerous flights of ducks were floating upon its bosom. Several running springs from all parts of its northern bank ran into the lake; and on its eastern side a large-sized creek evidently emptied itself into the lake, though the timber did not grow all the way down its channel. The water in this lake was eight or nine miles round—it was of an oblong form, whose greatest length is east and west. When quite full this basin must be at least twenty miles in circumference. I called this fine sheet of water Lake Wilson, after Samuel Wilson, Esq. of Horsham, Victoria, a munificent contributor to the fund for my expedition. There had been a disagreeable warm wind blowing all day. The position of this lake I made out to be in longitude 129° 52′.

Saturday, 27th September.—Yesterday had been exceedingly warm, at night the thermometer had not fallen below 80°, and by morning it stood at 76°; the hot wind from the north was still blowing. The hills of this range, from the shore of Lake Wilson, turned to the north-west, where the range ended in a few miles. I called the most western hill on the south side Mount Samuel, after Samuel Wilson, Esq., mentioned yesterday. Leaving Lake Wilson in its wildness and its beauty at the foot of its native mountains, I steered away to some low hills, to the south-south-west, which we reached in nine miles, where we found water in a creek, which I named Stevenson's Creek, after George Stevenson, Esq., of Melbourne, a contributor to my fund for travelling; we continued our journey for another hill, bearing 7° south of west; before reaching it however the country fell into numerous gullies, and at length formed the head of a creek, down which we went, and soon came upon water bubbling up from the ground below, and running down the channel, which was set with reeds and great quantities of enormous thistles. Some of the horses got bogged in this creek, as parts of the ground were very soft. The creek brought us out into a most beautiful amphitheatre, into which several creeks descended; it was covered with the richest carpet of verdure, and was a most charming spot. This valley or plain was nearly three miles across. We travelled over to its southern side, and camped under the hills, which surrounded it there. I found a sufficient supply of water in a very small creek-channel, and we encamped at it, having travelled only about fifteen miles on all courses. The grass and other herbage were splendid. The only opening to this beautiful oval was some distance to the east, consequently we must climb over some of the lowest hills in the morning to the south of us to get away.

Sunday, 28th September.—Thermometer last night 76°, this morning 78°. The night was exceedingly warm. Latitude of this camp 26° 8′ 11″. At starting this morning, we got over the hills to the south and came upon another fine valley, running westward, with a continuous line of ranges running parallel to the north of it. There were some hills to the south-west, and on steering towards them we struck Mr. Gosse's dray-track in about a mile. It was apparently going along this valley to the west. I however travelled south-west. At fifteen miles the hills were low, and mostly isolated, and there was but a poor likelihood of obtaining water at them. Other hills in a more continuous chain bore a little to the north of west, and we got into them in four miles. At three miles further we came to a valley with a green swamp in the middle. I found some water in it, but it was too boggy to allow horses to approach, so we pushed on for a round hill at the bottom of another valley, nearly in the same direction. We reached the hill late, and struck Mr. Gosse's dray-track again, and then we encamped; but we had no water, having travelled twenty-seven miles on various courses. The dray-track was now so much fresher, than formerly, that I intended to follow it, until I overtook Mr. Gosse, and I went to sleep dreaming, how I had met him in this wilderness, being a diluted parody upon "How I found Livingstone." The country passed over to-day was principally very fine valleys, richly clothed with grass and herbage.

Monday, 29th September.—Yesterday was rather a warm day. The thermometer at night stood at 82°, by morning it had fallen to 70°. Though we saw no water near our encampment on Mr. Gosse's track, yet in the night the horses found plenty in a small gully or creek channel about two miles back on the dray-track. I intended, as I said, to now follow this dray-track until I overtook Mr. Gosse, for I felt sure, that he could not now be very far in advance. But after following it for a mile, it turned suddenly to the south-west, down a valley with a creek in it, that lay in that direction. But as a more leading one ran also in a more westerly direction, at least it ran 22° north of west, I left the dray-track at right angles almost, and proceeded upon the more westerly line. The valley I now traversed soon became somewhat scrubby with mallee and triodia. There was a small watercourse with eucalypts, but quite dry, in this valley. In seven or eight miles we got much better country, lightly timbered with mulga and splendidly grassed. There were also some cotton- and saltbush-flats. At twelve miles the hills to our north receded, and there lay stretched before us a most beautiful plain, level as a billiard-table, and green as an emerald. Viewing it from the top of a hill, I could not help thinking what a glorious spot this would make for the display of cavalry manœuvres. In my mental eye I could see—

"The rush of squadrons sweeping,
Like whirlwinds o'er the plain,"

and mentally hear—

"The shouting of the slayers,
The screeching of the slain."

I called this splendid circle the Champ de Mars. It is, I dare say, fifteen or sixteen miles round, perhaps more. The hills on the northern side were much higher (we were on the south-east side) than those near us, there having been a continuous line of hills in that direction for a long distance back, and they seemed more inviting for water; so we rode across the field on a bearing of north 34° west to a kind of gully between two hills. At four and a-half miles we reached it, and found a rock-hole full of water up in a triodia-creek. It was almost entirely hidden amongst rocks and shrubs. It was seven or eight feet deep. The water drained into it from above. My horses lowered it considerably by the time all were satisfied. The day had been rather warm.

Tuesday, 30th September.—Last night was much cooler than the previous one, the thermometer at night standing at 62°, in the morning at 58°. When watering the horses this morning we found the rocky basin had been replenished during the night, and the horses not drinking much, we left it almost as full as we found it. A valley led away from here along the foot of the northern hills almost west. At four and a-half miles we crossed the channel of a fine little creek coming from them. It had several fine sheets of water. Most of the waterholes had rocky banks, and one pond in particular was of some depth. There was plenty of teatree about this creek, and it appeared to come a long distance through the hills from the north. The water was running; and I believe these are permanent springs. There were numerous families of ducks on these waters. The timber upon this creek was mostly bloodwood, of which we saw several fine specimens. Indeed, the bloodwood has now almost entirely supplanted the other eucalypts. There was another tree of a very peculiar leaf which we have frequently met before, but only as a bush. Here it had assumed the proportions of a tree. I collected some of its leaves for my kind friend Baron Mueller. This was a very pretty little spot. There were further several bare red hills festooned with cypress-pines, which gave a most pleasing tone to the scenery. I called these Harriet's Springs, after the christian name of my sister, Mrs. Henry Hull, of St. Kilda. The creek meandered away amongst the pine-hills, to the south-westward, and appeared to become much larger below where we were. At half-a-mile beyond this water I ascended a hill and found, that the two lines of hills encircling the Champ de Mars had now entirely separated, the space between becoming gradually broader. It was also filled with a good deal of mulga-scrub. A pointed hill at the far end of the southern line bore west from me, and we started away from it, entering scrub almost immediately. We continued on this west course for fifteen or sixteen miles, having the southern hills very close to our line of march. At that distance, having travelled twenty miles, I turned up a blind gully or water-channel, in a small triodia-valley, and found some water lying about amongst the grass, much to my surprise. The ground had a white sodaish appearance, and I followed it up about a quarter of a mile, where I found an excellent little pond full of water, at which we encamped. The water was not running, but I suppose it is supplied from below. This water lay under a rocky ledge, and if permanent or not I cannot say. The grass and herbage were splendid, and here we encamped; but the ants and burrs were very troublesome. We had been afflicted by the last-named vegetation upon several occasions, all through these regions. There was a high black-looking mountain on the northern line of ranges, which bore north-westward from here. It had a conic summit. I named it Mount Aloysius, in honour of the christian name of His Excellency the Governor of West Australia, whose remotest confines I have so lately entered.

Wednesday, 1st October.—Last night was cool and pleasant, the thermometer at night registering 60°, in the morning 52°. The latitude of this camp was 26° 9′. On leaving here this morning I continued upon the same line as yesterday, and towards the same hill, which was reached in five miles. It was almost the most westerly point of what had been the southern line of ranges, which we had been traversing. Climbing the hill, I found its summit was formed of great masses of rifted stone, which were either solid iron or coated thickly with it. The blocks rang with the sound of my iron-shod boots, while moving over them, with such a musical intonation and bell-like clang, that I named this the Bell-Rock. Mount Aloysius bore from here north 9° west, distant about ten miles. I here discovered, that it was quite an isolated range, as at its eastern extremity an open and level space could be seen between it and any other hills. There were a few low ridges to the south and south-westward. To the north-west and west-north-west the line of hills appeared to run on, though at diminished elevations; and masses of scrub were observable in almost all directions. Two hills standing together, with only a kind of gully or very small creek between, bore 11° to the north of west, distant ten or twelve miles away. I decided to go to them, and reached them in thirteen miles. Searching up the gully or creek, I could discover only one spot, where water appeared, amongst the stones high up the hill. There was evidently no great supply here, and I was afraid we should have to encamp without water; but I first took a ride down the channel, and saw where the natives had dug two wells; but now no water could be obtained there. I also noticed a bare rock, which I intended to visit on my return to the horses. Down the creek I found the channel perfectly dry; and on my return inspected the bare rock mentioned above. Close to it two or three natives had recently encamped, as their fire was just out; so I hastened to the rock, and there to my great satisfaction I found two fine rock-wells, of no great area, but of several feet in depth—one being much larger than the other, and containing quite a sufficient supply for all my horses. I soon called my party to the spot, and here we encamped. I called these very welcome basins Hogarth's Wells, after the Hon. Thomas Hogarth, of Smithfield, who most hospitably entertained Mr. Tietkens and myself, on our departure from Adelaide last year. The two hills I named respectively Mount Maria and Mount Jane, after Mr. Hogarth's two daughters—most agreeable young ladies—whom I met at their father's house.

Thursday, 2nd October.—Thermometer last night 78°; this morning 68°. I was sorry to be compelled to leave one of these cisterns empty, which for ages the simple inhabitants of these regions had probably never seen dry before, as for all their requirements I should imagine water was always obtainable there. From here a few low-timbered ridges ran south-westerly, and others more to the west, but they also seemed very low. To the north-west and west-north-west they also seemed to run out entirely. Having Mr. Gosse's dray-track on my left or south of me, I thought that by going to the south-westerly hill I might intersect it; for not having seen it for some time I felt sure that

I must be in close proximity to him, if indeed I had not already passed him. It was late before we got away from this camp, reaching some ridges in nine miles, which lay south 55° west. There was no water to be found in these hills. Further south were others of a similar kind some ten or twelve miles away. Beyond them the country appeared all scrub; scrubs also intervened between this hill and any others in all directions. The western horizon was bounded by broken ranges with some high points amongst them; they were a long way off—the country being all scrub between. To the north-west and to west-north-west some bald ranges ran on. I was sorry I had not followed them instead of coming here. I therefore made over to them, steering for a fall or broken gap in them to the north-north-west, which was reached in nine miles. In a glen I found a watercourse with plenty of teatree growing on its banks. Further up, the channel was completely choked with good-sized teatrees, whose branches, entirely interwoven in one another, prevented any animal larger than a man from approaching the water, which bubbled along at their feet. We got here rather late, and had to chop away a quantity of the timber, which grew so closely together. These hills are perfectly bald of timber, and consist of exposed masses of rifted granite, which could only have become so by the action of earthquakes, which at one time or other must have been exceedingly frequent in this region. I ought indeed to have mentioned it previously, that the hills and mountains of the Mann-Range, some few of the more westerly of the Musgrave-Chain, and all west of the Mann-Ranges, have been shivered into fragments by volcanic force; and most of the higher points of all these hills are composed of frowning masses of black-looking or intensely red ironstone or granite coated with iron. Triodia grows as far up the sides as it is possible to obtain any soil, but even this plant cannot exist upon solid rock, therefore all the summits of these hills are bare. These shivered masses of stone, having large interstices amongst them, are the resort of countless swarms of rock wallaby, which come down upon the lower ground at night to feed. These hills all have the metallic clang of the bell-metal rock, and are all highly magnetic. Soon after starting this morning Gibson found a lowan's or scrub-pheasant's nest, and we obtained six fresh eggs. Subsequently I found another in the scrub to-day, from which we obtained five more. We saw several native huts in the scrub to-day, which were of very large dimensions, having the largest trees they could get to build them with. Where they got their water, when living here, is a mystery, unless they obtain it from the roots of the mulga-tree, which must be the case, as we occasionally see small circular pits dug up at the foot of some trees, which generally die after the operation of tapping.

Friday, 3rd October.—I gave the horses a day's rest here to-day, as I had a great many matters to attend to. I called this little spot, where we had an excellent camp, Glen Osborne, after Samuel Osborne, Esq., of Elsternwick, near Melbourne. The day was exceedingly warm, the thermometer at night not having fallen below 90°, and by the morning of

Saturday, 4th October, it stood at 76°; another day's rest or so-called rest was required, which with my party usually means a hard day's work, it being only a respite for the horses; we continually have so much sewing and repairing of packbags, that any one unacquainted with the fact by experience can perhaps scarcely conceive it.

Sunday, 5th October.—Last night was delightfully cool, the thermometer standing at 68°, and by morning at 46°, quite a change having taken place in the temperature of the atmosphere. Having watered the horses and packed them all, we started away for some other hills that lay in a south-westerly direction; in seven miles we passed the foot of a low range, and thence on to the higher hills; we reached them in twenty-two miles from Glen Osborne. A portion of the country traversed by us was well-grassed mulga-scrub, but not thick; on nearing these hills we passed over some fine cotton-bush flats, and immediately at the foot of the hills we found ourselves upon a very fine piece of plain, beautifully grassed, and liable at times to inundation. I could find no water in the first valley we reached, and as it was late we had to encamp without any; the day having been delightfully cool, the night also promised to be cold; so I did not feel much anxiety on account of the horses. These hills are similar to those previously described in this journal, being greatly impregnated with iron, and having vast upheavals of iron-coated granite, broken and lying in masses of black and pointed rocks upon all the summits and down their sides; they were all highly magnetic.

Monday, 6th October.—Last night was again delightfully cool, the thermometer at night standing at 68°, in the morning at 48°. As there was not any water in the camp I climbed over the hills near; from the top I saw a teatree-creek in a glen below, which promised water; it was some distance off, two miles at least; I however walked down to it and found a strong running stream. Returning to the camp, we started away over some of the lower hills, intending to go where I had found the water; but we found some more and an easier road for horses over the hills. This creek ran southwards, the other I had found ran west; we followed the southern channel, which ceased to contain water in a mile, and ended entirely where the hills fell off, a piece of fine open grassy country surrounding these ranges. At five miles from camp, intending to go to some hills south-westerly from here, I again struck upon Mr. Gosse's dray-track, and was certainly startled to perceive, that the vehicle had now returned upon its old track; the tracks seemed reasonably fresh, but it is difficult to tell within a week the date of such marks. I could only guess, that Mr. Gosse had not penetrated far beyond this range, and that he must now have decided to return to South Australia. I had for some days anticipated meeting him, but he now having returned I did not follow after him, but followed the dray-track on its outward course; it brought us to where it crossed the channel of the same creek I had walked over to this morning, and though there was no surface-water, we obtained it easily in the sand. A peculiarity of this region seems to be, that water cannot exist far from the rocky foundations of the hills; the instant the valleys open and any soil appears, the water sinks, though there may be a fine stream running just above. The country all round these hills is very fine; there is a little scrub here and there, but it is not thick, and the whole region is excellently grassed. I believe the water, which I saw this morning up this creek, is permanent (and I had reasons afterwards for still believing so, as I had to retreat to it twice, and eventually called our encampment here Fort Mueller).

Tuesday, 7th October.—Last night was again most agreeably cool, allowing me the indulgence of a full supply of blankets. The thermometer at night stood at 70°; this morning at 50°. I found my position here last night to be in lat. 26°. 12′, long. 128° 0′ 0″. The camp was not moved to day, but Mr. Tietkens and I followed the dray-track, which led us in about two miles to where Mr. Gosse's camp had been. It appeared to me a most wretched hole, being in a thick scrubby gully, with a small teatree watercourse in it; there were several shady gunyahs and grass beds, where the several members of the party had slept. When Mr. Gosse left or since, the water had dried up; but there was still a small quantity further up. No valuables of any kind had been left by the party; one or two pieces of light trace-chain were hanging on a tree. A bloodwood-tree was branded G O S with 14 under it; but that conveyed no information whatever to me. There was a pumpkin-vine growing in the creek, but it looked very sickly indeed. We returned to our camp: the dray-track went no further than this place.

Wednesday, 8th October.—Leaving this encampment, I struck away for a new line of ranges, whose highest point bore 14° to the south of west. We only got fifteen miles this evening. The country travelled over was very peculiar, and different to any we had previously met with; it was open, covered with tall triodia, and was mostly composed of limestone. At intervals numbers of eucalypts appeared, also some bloodwood-trees and native-poplars. There was no grass for several miles, and I hardly expected to find any for the horses; but we found a small patch of scrub, in which there was some. The ground all around was stony and covered with triodia. In our travels to-day we stumbled across three lowan's nests. The first, though of enormous size, produced only two eggs; the next gave four, and the third gave six. We thanked Providence for supplying us with such luxuries in such a wilderness; whether the lowans had equal reason for thanks may be very reasonably doubted. There are much easier matters to perform in the world, than the carrying of lowan's eggs. For those of my readers, who do not know what they are like, I may inform, that they are as large, if not larger, than a goose-egg, but of a more delicious flavour than any other egg in the world; their shell is so fragile, that if a person is not careful even in lifting them the fingers will crunch through the pink-tinted shell in an instant; therefore carrying a dozen of such eggs is no easy matter. I took upon myself the responsibility of bringing them safe into camp, which I accomplished by packing them in grass tied up in a handkerchief and slung round my neck, forming a fine fardel on my chest and immediately under my chin. A photogram of a person with such an appendage would scarcely lead to recognition. We used some of the eggs in our tea as a substitute for milk, which is an agreeable change in the bush. A few of the eggs proved to possess some slight germs of vitality, the preliminary process being in the formation of eyes; but explorers (in the field) are not such particular mortals, as to stand very nice upon such trifles—indeed parboiled youthful lowans' eyes are considered quite a delicacy in the camp.

Thursday, 9th October.—Last night the thermometer stood at 68°; this morning it had risen to 70°. At early dawn I noticed lightnings to the west, and soon the whole horizon became cloudy in that direction. Thunder was afterwards heard, but what storm there was passed away to the south of us. Making an early start (as the horses did not leave the grass patch, upon which they were put last night) we proceeded on our course; the limestone gradually disappeared and sandhills took their place. At twelve miles we reached two low ridges thickly timbered;

the main-hills, for which we had been steering, were still five or six miles distant, the intermediate space being filled with dense scrub; one or two pine-clad hills on our right with bare rocks attracted my attention, and I decided to visit them first and search for water. There were no such things as rock-holes, or any place where water lodged. While I was searching, Mr. Tietkens found a native well under the foot of one of the pine-hills; it was rather deep, with water at the bottom; it being dug out of a kind of limestone. As it would take some time and a great amount of work to supply from it water to all the horses, I thought as the higher hills were more likely to have some surface-water about them, it would save time to push on to them. So we did not trouble the native well; nearing the hills we saw some of Mr. Gosse's horse-tracks, both coming and going. On searching about amongst the scrubby gullies of the main-hills, I found a supply of water, though there was no great quantity; but it was running with but a very weakly current. Mr. Tietkens found a better supply in another channel a mile further north. There was every appearance of approaching rain; we got everything under canvas. We had travelled seventeen miles; I was in great hopes of rain falling, but was disappointed.

Friday, 10th October.—A heavy gale of wind sprang up at night, and blew away any rain, which might have been expected to fall; but as it was still cloudy, we remained here to-day. Mr. Tietkens and I climbed the highest of these hills, which was thickly covered with mulga and scrubby bushes. From its summit the appearance of the surrounding country was most peculiar and strange; to the west and round by N.W. to N., was a mass of broken timbered hills with scrubby belts between. The atmosphere was too hazy, to allow of a very distinct vision; but I could distinguish lines of hills, if not ranges, to the west or a little south of it for a long distance. The appearance of this country was certainly not the most gratifying; but as the hills still run on, though entirely different now from those behind us, our only hope is, that water will still be discovered in them. Close to the spot where I reached the summit, I observed a small pile of stones erected, I suppose by Mr. Gosse as a memorial of his visit, or as a point for triangulation. I think the view from this hill to a traveller, in any way predisposed to retrace his steps, would at once put a stop to any further progress westward; for the whole region round about was enveloped in thick scrubs, and the hills which were visible about it may be said to have been only just so.

Saturday, 11th October.—The sky had remained cloudy all yesterday, and I hoped, if the wind would only cease, some rain would surely fall; so we remained here again to-day hoping against hope. For though we had most powerful reverberations of thunder, and forked and vivid lightning played in all parts of the sky, yet no rain would fall, though the atmosphere was charged with moisture and electricity all round.

Sunday, 12th October.—This morning we had nearly three minutes' sprinkling of rain, when the sky became perfectly clear. Thermometer at night stood at 64°; this morning at 58°. At starting this morning, we had to go round the hills to the south for three or four miles, then struck west through a mass of hideous scrubby gullies and over ridges. The scrubs being so thick, two or three horses got clean away from us three different times, and it caused me a great deal of anxiety and work, to find them again. After turning to the west through the hills, I continued on that course, until we struck a small creek with one or two eucalypts in it; in its bed I found a small native well, in which the water was very near the surface; we therefore encamped, having travelled ten or eleven miles, but only making six miles good S.W. We easily obtained plenty of water for all our horses, and as there was a small patch of open ground along the banks of the little creek, it was a very good, not to say pretty encampment; the grass and herbage here were exceedingly good and beautifully green. We slept upon the sandy bed of the creek to escape the terrible quantities of burrs, which grew all over these wilds.

Monday, 13th October.—Last night was certainly not very cold, the temperature in the morning being 2° above that at night. Starting this morning, I steered for the highest hills I had seen from Gosse's Pile lying westerly. There appeared a gap and a fall between two hills. We only reached their foot at night, when we encamped upon a small dry creek, having travelled twenty-two miles. The country passed through to-day was all scrubby and exceedingly thick, which was fully proved by the number of pack-bags that got torn, and which we had to mend at a late hour.

Tuesday, 14th October.—The thermometer last night stood at 64°; in the morning 68°, indicating the approach of a warm, if not a hot day. Getting under weigh we proceeded up the densely scrubby and stony valley, and on reaching the top, I was disappointed to find, that though there was an open valley below the hills, all round seemed too much disconnected to form any good watering-places; however we descended into the valley below, and leaving Gibson and Jimmy with the horses, Mr. Tietkens and I rode away in different directions, to search for water. In about two hours after parting we both met in the only likely-looking little watercourse seen by us, which we followed up, and found a most welcome and unexpectedly large pond (for such a place) at the foot of some rocks, which above the pond had some rocky basins, which contained water, with rather a pronounced odour of stagnation about them; above them again the water was running, but there being a space between, which absorbed it from the surface, we returned and brought the horses as near as we could find a convenient spot and encamped. We had not travelled more than five or six miles, but I considered it a very good day's work. The valley, in which we encamped, ran north-east and south-west; it was very narrow and not altogether too open, nor was the grass too good, the greater portion being covered with triodia. About eight o'clock at night a thunderstorm came up from the west and north-west, only sprinkling us with a few drops of rain; it proceeded round by north to east, thence to south, and performed a complete circle, reaching its original starting point in about an hour; it then travelled to the northwards again, and finally disappeared, the remainder of the night being beautifully calm and clear.

Wednesday, 15th October.—Thermometer during the thunderstorm last night stood at 70°; by morning it was down to 58°. Some horses required shoeing for the first time, since leaving the Telegraph-Line. From the top of the hill, above the camp, Gosse's Pile bore 15°. to the south of east; the horizon to the west was bounded by low scrubby ridges, with an odd hill here and there cropping out above them, but they were all very low. One hill, as high as most of the others, bore 23° to the north of west, and I determined at starting to steer for it. There was another line of ridges a little more west and a trifle south of it, and they seemed a thought higher than those I intended to reach; there was very little to choose between them; hills also ran along northward and north-eastward for some distance. At eight o'clock again to-night another thunderstorm came up from a more westerly quarter, than that of last night; it sprinkled us with a few drops of rain, and became dispersed to the south and south-east.

Thursday, 16th October.—Thermometer last night stood at 74°; by morning at 58°. We passed the day in shoeing horses, mending bags and general repairs. The weather was most delightful. While out after the horses, Mr. Tietkens found another supply of water, about two miles from the camp, southward, and on the opposite side of the valley.

Friday, 17th October.—Last night was delightfully cool, the thermometer at night indicating 58°, and 48° in the morning. Finishing what work we had to perform we remained here again to-day. I found the water boiled in this valley at 209°, making the approximate altitude of this country 1534 feet above the sea-level.

Saturday, 18th October.—Thermometer last night 58°; this morning 50°. It was late, when we got away from the litttle valley, which I called the Shoeing-Camp. We had remained at it longer, than at any other encampment since I started. Getting over a low fall in the hills opposite the camp, we turned upon our proper course and travelled fifteen miles, the first two or three being through very fine country, well grassed, and having a good deal of saltbush, being lightly timbered and no triodia upon it. However the scrub and triodia very soon made their appearance together, and where we had to camp we found but a miserable place for the horses, there being neither grass nor water.

Sunday, 19th October.—Thermometer last night 58°; this morning 46°. Getting an early start, we entered on our yesterday's course; but seeing a bare-looking rocky hill a little to the north of our line, I made up to it, reaching it in ten miles. On searching about I found several small holes or cups worn into the solid rock; and as they mostly contained water, I had the horses unpacked, while a further search was made. Leaving Gibson and Jimmy with the camp, Mr. Tietkens and I rode away in different directions. I went away to the hills, for which I had been steering, and searched among ridges, rocks and gullies. I found no water however. I next travelled down a pretty little valley, beautifully grassed, with the clay-bed of a small watercourse stretching west. Following it down some eucalypts appeared upon it, and I was in great hopes of finding water. There was a considerable quantity of salt in some of the little holes, which had only lately become dry. In advance of me I saw the valley was closed by hills facing me. This little channel emptied itself into a very pretty eucalyptus creek; there was also another channel opposite, which emptied itself into the main-creek. I began to be in great hopes of soon falling upon a considerable water-supply. Upon the banks of this creek I found growing, for the first time on this expedition, great quantities of that gorgeous flower, the Desert or as it is usually called, Sturt's Pea; not that that celebrated traveller was the discoverer of it. I followed the main-creek down; it ran south-west. The ground near the banks and channel was greatly impregnated with salt, and I concluded to find only brackish water here. In about a mile I saw a small sheet of water ahead of me, under a rocky bank. On reaching it I soon saw, that it was none of the best; on tasting it I found it salter than brine. My horse was most anxious to get at it. I washed my face and head in this pond, and the salt settled in bright glittering particles on my hair and beard. I went some distance further down the creek, and saw several places that had only just dried up; but the glittering salt left at the bottom told only too truly of what quality the lately absorbed fluid had been. The creek seemed to meander right through these ridges, which hedged it in pretty close, and sent numerous small branch channels into it. I left the main-creek, and searched among those, but quite unsuccessfully. After some hours of ineffectual search amongst the hills fringing the creek and branch-creeks running into it, I determined to try the main-channel above where I had first struck it. About a mile above that spot I found a place, where the sand in the bed was damp. I set to work, and scratched at it with my hands, making a hole two feet deep, when I obtained the element I was so anxiously searching for, but upon tasting it I was discouraged to find, that though not salt it was more bitter than gall, so much so, that it nearly made me sick when I swallowed a few drops of it. My horse would only take one sup, and that was sufficient for him also. I was greatly perplexed what to do; there was no water at the camp in the cups for all the horses. I had found none, and it was now sundown. I sincerely hoped Mr. Tietkens had been more successful. I was compelled to return to the camp, which I reached late. Here I heard that Mr. Tietkens had found a few more cups at another hill, and that twelve of the horses had had a partial drink, there being but a small quantity left for the others in the morning. It appeared also, that when Mr. Tietkens and Gibson took the twelve horses away to the other hill, and Jimmy Andrews was alone in the camp, he walked over to a little cup reserved by us for our own use, to fill a tin-billy, to make some tea; walking unconsciously along with his eyes on the ground until he approached the cup, when to his horror and amazement he discovered, that some thirty or forty aboriginals were seated or standing round the spot. When he neared them, they yelled in chorus, which also elicited a yell from him in return, and letting fall the tin-utensils which he carried, fairly ran back to the camp, when he proceeded to get all the guns and rifles, &c., in readiness to shoot the whole lot, if they came to attack him, but Mr. Tietkens and Gibson returning with the horses (having heard the yelling at this moment) caused the natives to decamp, and relieved Jimmy's mind of its superabundant load of care and fear. These poor creatures were no doubt dreadfully annoyed to find their little reservoirs discovered by such water-swallowing wretches as they doubtless thought white men and their horses are. I could only console my mind with the reflection, that in such a region as this water must be taken when found at any price; but I dare say, they knew where to get more, and I did not. I fancied some natives were prowling round our encampment during the early part of the night, as my little dog kept up an almost incessant barking for two or three hours; after that he became quiet.

Monday, 20th October.—The thermometer last night stood at 60°, this morning at 50°; the night was agreeably cool. On waking this morning I determined, to move the horses over to the bitter water found yesterday, thinking that if a quantity were baled out of the tank the quality might become improved. There was only a couple of buckets each left for the twelve horses that had no water yesterday, so after emptying every one of these little cups we departed. The bitter water was about six miles from the cups. Mr. Tietkens went on in advance of us on my tracks of yesterday to search for water, while I took the horses to the bitter water. On reaching it we found, that though all the horses crowded eagerly round the tank which we dug, none of them would swallow more than a mouthful at a time. We left them to their own devices, while we tried several places further up the channel, but in every instance, though we always obtained water, it was either bitter or salt. I then left the camp, and rode away to the west amongst the hills and renewed my search, but was quite unsuccessful; there was literally no water to find. On mounting one of these hills I found, that they continued for several miles to the west. They were densely covered with shrubs, and composed of red granite rocks, all riven, and standing up edgewise, making it exceedingly difficult for a horse to traverse them. To the north the horizon was bounded by a low scrubby ridge. I continued my search all the rest of the day, in all directions, amongst these detestable hills and gullies, going wherever I could see a fall or valley or break in the hills; but by night I had not found a drop of water. I returned to the camp, hoping to hear that Mr. Tietkens had been more fortunate. I was in a very gloomy frame of mind. Most of the horses were still standing at the bitter tank, looking hopelessly into it for purer water. There was splendid grass and herbage at this camp, so I drove all the horses away, as it only made them the more thirsty, to stand moping over their miseries. Mr. Tietkens had not returned, and I awaited his arrival with considerable anxiety, as if he had been equally unfortunate as myself, I should have no choice but to retreat to the Shoeing-Camp. There was no object or feature in any direction to which to make from here, except indeed, to follow the creek, for where it could run to was certainly a mystery, and it did not promise very well, as I had found no water in it as far as I had travelled down it. It was after dark when Mr. Tietkens returned, when he informed me, that all his searchings had been unsuccessful, until returning through the hills he stumbled across a small eucalyptus-creek, where he had obtained enough water for his horse, by removing a quantity of stones, and probably by our digging out a good-sized tank we might get one drink for all the horses. This was a great relief to my mind, though it was not too promising. The day had been exceedingly warm, the thermometer had gone up to 98° in the shade. Mr. Tietkens also informed me, that he had gone some distance down the main-creek; he had found it perfectly dry, and had abandoned it in despair at last, though it had grown larger. The place where he had found the only water was about three miles away from here.

Tuesday, 21st October.—The thermometer last night stood at 60°, this morning at 58°; the night was pleasantly cool. We got the horses early, and moved down the main-creek three miles, to where the little creek with water found by Mr. Tietkens joined it; it was over a mile up it to the water. Mr. T. and Gibson took the shovel and went to open out a tank, while Jimmy and I unpacked the horses. In about an hour Mr. Tietkens came down and took eight horses up; but it was past mid-day before they got a drink sufficient to satisfy them; then eight more were taken. I went up with them, and found that the supply was so limited, as there was only a very slight drainage, that it was nearly evening before the second eight had been watered, and the supply diminished so that the third eight could not get a bucketful each by night, but as a large hole had been made, I hoped to see it full in the morning, though I was in a most unenviable frame of mind, because the thought of retreating was as bitter as the taste of water in the creek.

Wednesday, 22nd October.—The thermometer stood last night at 74°, by morning at 58°. To-day was spent by Mr. Tietkens and Gibson doling out a few quarts of water only to the horses—those that had none yesterday getting as much as could be given; very little had drained out into the tank during the night, and I decided to be off at once; but Mr. Tietkens advised a further search. I spent the day in fruitless searchings for the fluid, which I was evidently not destined to discover, if indeed it existed at all here, and this I very much doubted. Six weeks or two months ago there must have been plenty of water here; and could I have only been here at that time, I have no doubt I should have passed across to the Murchison-River; but as there was literally none here now, what was to be done? When I again returned from an unsuccessful search, I found that not half the horses had yet received their miserable allowance; they were now all huddled up round the rocks, where the scanty supply came from. One little wretched cob was upon the last verge of existence; he was not well, had been falling away for some time, was now as poor as a crow, he had not carried anything but a packsaddle for some time, and was always lagging behind and planting himself in the scrub, which had often caused me great annoyance; so as he was already nearly dead, and we wanted some hide, and there being no water for him, I dispatched him from this scene of woe; there was not sufficient flesh on his bones to cure; but we got a quantity of what there was, and because we fried it we called it steaks, and because we called it steaks we said we enjoyed it, though it was utterly tasteless; the hide was quite rotten and useless, being as thin and flimsy as brown paper. The horse ridden all the day by me could get no water when I returned, and I thought it high time to come to some decision as to what to do. It was useless to push hap-hazard out farther west from here before getting the horses a drink, as they were now all as thirsty as ever; so my only alternative was a retreat to the Shoeing-Camp, though we could not well reach it in a day. To-day was very warm, the thermometer in the shade going up to 99°.

Thursday, 23rd October.—The thermometer last night stood at 74°, this morning at 54°. The horses had split up into two or three mobs, and it was late before they were all mustered. Only a quart or two of water had percolated into the tank during the night, and that we took with us for ourselves. The horses all looked exceedingly hollow, but when put upon our outcoming tracks they stepped out with a will. We had some difficulty in getting them past the cups, as some of them ran up the rocks to where they had formerly received water, and of course they expected it again and probably supposed it always remains there; but that was a fallacy, which we were compelled somewhat rudely to induce them to understand. Some of these peculiar little indents held only a few pints of water, others a few quarts, and the largest only a few gallons. We pushed on until dark and encamped again without water, having travelled twenty-five miles. We short-hobbled the horses, but were greatly afraid, from their apparent eagerness to continue on the tracks, that they would stampede in the night. I should have continued on until we reached our old camp, had the country been otherwise than scrub.

Friday, 24th October.—Yesterday had been quite warm enough for my personal convenience; the thermometer had indicated 101° in the shade; the night also was close and sultry, the thermometer standing at 80° at night and 76° in the morning. Mr. Tietkens and I were upon the tracks at the first glimpse of light, and though we soon overtook some of the horses we found that the greater number had gone some distance. At three miles we found nine more. It took us close to our old camp-valley before we came upon the remainder. As soon as we unhobbled them they tried to give us the slip; we were therefore compelled to drive them back to the next mob in hobbles, for the country being so scrubby that a rider bare-backed has but a poor chance against a horse determined to go his own road. When we started we had no occasion to drive them, but rode ahead, and they followed, as they well knew, where they were going. We had left but a small supply of water here, when we left this place, but Mr. Tietkens having found some more I did not anticipate any trouble. Mr. Tietkens went on in advance of us, to see how it was holding out. Six days having elapsed makes a great difference in water, that is inclined to depart. When he joined us again he advised, that we should go to the new place, as he believed very little digging would produce a sufficient supply. Mr. T. and Gibson went up with the shovel, and Jimmy and I unpacked the horses. They all walked off up the gully after getting the packs off. They had some hours to wait however, before they could all be satisfied. In the meantime Jimmy Andrews set fire to the grass close to all the packs, and a strong hot wind blowing soon placed them in jeopardy. The grass was dry and thick, and the fire raged around us in a terrific manner. Guns and rifles, riding and pack-saddles, were surrounded by flames in an instant. We ran and halloaed, and frantically threw them out on to ground already burnt. Upsetting a couple of packs we got the bags to dash out the flames with. It was only by the most desperate exertions, that we saved almost everything, and the grass where the things had been, got burnt as soon as ever they were removed, the spot from which each article was lifted being instantly on fire. I was on fire, Jimmy was on fire, my head and brain were in one whirling blaze: and what with the heat and dust, smoke, ashes and wind, I thought I must have become suddenly translated to Pandemonium, and our appearance was certainly most satanic, for we were both as black as demons. The day also had been extremely warm; the thermometer had stood at 105° in the shade at the camp; but I think a degree or two extra was owing to the heat in the valley caused by the fire, which was still raging and which continued until night.

Saturday, 25th October.—Last night was close and hot, the thermometer standing at 80°, in the morning at 68°. We had left all the horses up in the gully last night, as just above the watering-place was an open piece of grassy ground. The water-supply here not being so good as we expected, I determined to try what effect digging a tank at our old place would have; so we packed up and went off to it. By digging we got a sufficiency of water, and here we again encamped. The camp will have to remain here for a few days, while I take a tour round the country in search of more water. The thermometer to-day stood at 100° in the shade.

Sunday, 26th October.—Thermometer last night 80°, this morning 66°. The night very warm until near morning. To-day being Sunday we indulged in a real day's rest, and were able to obtain a kind of bath. The horses are running at large in the valley, and will, I hope, soon recover from their late deprivations.

Monday, 27th October.—Thermometer last night 76°, this morning 58°. I had determined since having had to retreat from the west, to try a course more northerly, in hopes of discovering some hills or ranges in that direction. Mr. Tietkens and I therefore got three horses, and taking a week's supply of provisions, we started away for the north, leaving Gibson and Jimmy behind at the camp. I had some doubt on my mind as to the water lasting while we were absent, as it only drained in from a small gully. After leaving the camp we had the hills of these ridges before us for three or four miles, but after crossing them we had the usual triodia-country, consisting of sandhills and occasional mulga-scrub flats. For twenty miles the scrubs were not thick, and I began to hope a more open country was before us; but we found, that such was not the case. From a little stony rise, at twenty-one miles, we could see some ridges ahead, and I hoped to obtain water in them. The day was sufficiently warm, the thermometer at 101° in the shade at one o'clock, when we turned our horses out for an hour, starting again at 2 o'clock, during which time the thermometer rose 2°. We continued our course, and reached the ridges in thirty-nine miles from the camp; we looked about for water, but these hills were of too trifling an elevation to have any drainage—from them, nothing but scrub could be seen to the north; to the east of us a line of hills, connected with those at the Shoeing-Camp, had been running parallel with us, but were now shut out from view by a scrubby rise. We encamped here without water.

Tuesday, 28th October.—Last night became suddenly cool, the thermometer at 9 p.m. went down to 66°; by morning it stood at 60°. We got away from camp at an early hour, and after visiting one or two more of the hills near, we gave up the search for water as hopeless. We proceeded still north; at five miles we arrived at a scrubby sand-ridge, and obtained a most displeasing view of the country further on; it consisted of a more depressed surface, entirely filled up with dense scrub, with another ridge similar to the one we were on bounding the view. We entered the scrub, and in eight or nine miles came to the ridge; we had to travel a mile and a-half through it before we obtained another view, which, when got, was precisely similar to that behind us, that is to say, scrubs and a ridge, but now the scrub was much more extensive, the next ridge being fifteen or eighteen miles away. We had come fifty-one miles from the camp; there was no probability of getting water in such a region. To the west the horizon was bounded by what appeared a perfectly flat and level line, which ran up northward as far as could be seen. This horizon seemed not more than twenty-five to thirty miles away; between us and it were a few low stony hills. Thinking it useless to go further north through the dense scrub which existed there, I determined to push over to the western line, which looked so flat. I said before, there were one or two stony hills in that direction; we reached the first in twenty miles, the next was formed of nearly bare rock, and seeing some native gunyahs about, upon searching we found another of those extraordinary little basins or cups, washed out of the solid rock by old ocean's force, ages before, an all-seeing Providence placed his dusky children upon this scene, or even before the waters had sufficiently subsided to permit either man or animal to exist here; and in this singular cup we obtained a small supply of that fluid, so terribly scarce in this region. We had to give it to our horses in a canvas bucket, and we camped at this exceedingly welcome spot. There were a few hundred acres of excellent grass land about, and the horses did remarkably well for the night. The horses drank at least half the water that was in the cup, and it was evident that we should leave but little behind us. It was early when we reached this singular and not unpleasing little spot, so the horses had time to get a good feed. We encamped here, having travelled thirty-three miles. From the top of these ridges nothing but scrubs appeared to the north, while to the west the horizon was now not quite so flat as it had formerly appeared, but now seemed to consist of low undulating ridges. The day had been very hot, and the horses were very thirsty when we reached here; the thermometer stood at 106° in the shade.

Wednesday, 29th October.—Last night was exceedingly hot and close, the thermometer at 8.30 p.m. standing at 90°. This proved to be a most abominable camp, as we could get no sleep all night, the ants were so frightfully numerous and troublesome, and kept biting us so perpetually that we were compelled to keep moving about until morning. The sky became greatly overcast last night, and a few heat-drops of rain fell; we were only too glad to leave this wretched camp; the horses came up themselves very early for more water, and required so much, that there was scarcely a quart of fluid left in the cup when we departed. Starting at six o'clock, we continued on our west course, over sandhills and through scrub and spinifex; we reached the low ridges of which the western horizon was composed in five miles; no view could be obtained from them; a mile off was a slightly higher point, which we went to, when it appeared that this kind of country ran north and west, without any other object upon which to rest the eye; there were a few little gullies, which we only wasted an hour for searching through, for of course there was no water. I was not at all satisfied in my mind at the thought of our retreat from the bitter-water creek, which now lay south of us. I was anxious to find out, where it went to; for though we had spent several days in its neighbourhood, yet we had not travelled more than eight or ten miles down it. I resolved to go there again, as we might still expect a bucket or two of water, where I had killed the little cob, and which place we always called the Cob. I therefore turned to the south in hopes, that I might yet get some satisfaction out of that region at last, for a creek in this part of the country in any way independent should not be lost sight of. We were now some 38 or 39 miles from our old camp, or 41 from the Cob-Water. We departed towards it again south. All this while I was exceedingly anxious about the water at the Shoeing-Camp. At 25 miles from our last night's camp we turned the horses out for an hour; the day was exceedingly warm, the thermometer again standing at 106° in the shade, though the morning had been cloudy;

it was late enough in the evening when we reached the old tank, and the quantity of water that had accumulated since we left—amounted to nearly a bucketful. I was greatly disappointed; I only asked three buckets, but there was scarcely one; it was evident the drainage was fast ceasing.

Thursday, 30th October.—The thermometer last night again stood at 90°, by morning at 70°; the sky was overcast and threatened rain during the night, but none whatever fell. We found the three horses huddled up together at the water-tank this morning, they probably having been there all night; the tank was quite empty, when we got the horses. I started down the creek, one or two small branches joining it made the bed larger at eight miles from the Cob. We ascended a low hill and saw the creek-timber five or six miles to the south, the country on all sides being thick scrub. On reaching it I thought it was a different creek, but it ran in the same direction, it was much larger, but still no water in its bed anywhere; we also found an old native well, which we scratched out with our hands, but it was perfectly dry. At twelve miles we came to the junction of another creek with this one, the new one coming from some hills to the east, and immediately below the junction the bed became filled with green rushes; we had great hopes of getting water here; we had been compelled to leave the shovel with Gibson at the Shoeing-Camp, or we could have dug out a hole here, but the ground was too hard to do it with our hands. We turned the horses out here for an hour. Starting again and following the creek, another and much larger creek came from the east and joined this one, or rather our creek ran into the other one; there were some large holes in the new creek, but all dry; we first went some distance up it and then turned to some low hills a mile from its southern bank, but others further away intercepted the view, and we could only see the creek-timber for a short distance. We had now a range of low hills to the east, from which several small channels had come down. Mr. Tietkens desired to search among the hills to the east of us, while I more inclined to the creek; however our horses were now in a bad state, and it was necessary to think of them, so I decided to return to the Shoeing-Camp, and search the ranges east of us on our road, but still intended if we got water to return and follow the creek, which I conceived might yet lead to the discovery of some important geographical features. It was night by the time we had reached the foot of the hills—we had travelled a long way and our horses were very thirsty; we found no water where we struck the hills. We were now nearly west from our Shoeing-Camp, the only place in this wide-spread desert, where we knew that any water existed, and we could not be sure if it yet existed there; we were yet thirty miles from it, but hoped to find some in the hills next morning. The day was not quite so warm as yesterday and the day before, the thermometer only rising to 100° at one o'clock, which is not quite the hottest time of day; from 2.30 to 3 p.m. being the culminating hour.

Friday, 31st October.—The temperature changed considerably yesterday evening, and the thermometer at night was down to 60°, in the morning 50°; the night appeared to me quite cold. The horses seemed in great distress this morning; we soon got into the hills and searched about in creek-channels, gullies and wretched places, until nearly midday, without finding any water; so I resolved to depart at once as the horses were now very bad, having had only one-third of a bucket of water each since Wednesday. It was late enough at night when we arrived at the camp, the horses being scarcely able to drag themselves along. Fortunately the day had been remarkably cool, the thermometer not indicating higher than 80° in the shade. On reaching the camp I was truly gratified to hear, that the water had not only held out well, but Gibson declared it permanent; however I knew better, and I had hurried my movements back to the camp in consequence of my anxiety with respect to it. I had also been in great bodily pain from an attack of illness, which made riding most agonising, or I might have made a more strenuous effort to reach a further northern point soon after starting.

Saturday, 1st November.—The thermometer last night went down to 46°, and by morning it was actually as low as 32°, though there was no ice. I do not remember ever feeling such a sensation of intense cold. We remained in camp to-day. Thermometer 85°.

Sunday, 2nd November.—Thermometer last night 54°, this morning 50°. Another splendid cool day, and being Sunday we thoroughly enjoyed a day's rest in the camp.

Monday, 3rd November.—Thermometer last night stood at 60°; it was also at 60° this morning. I was exceedingly anxious to go back to the junction of the two creeks and follow the channel further; but I could not ride for a few days, and I got Mr. Tietkens to take Gibson and the shovel, and go there again; first to where we had seen the green rushes, and try if water could be got by digging, and if not, to return up the creek, which joins close to the rushes, and search the hills also about there. They got away at a reasonably early hour. The day was warm enough; the thermometer stood at 100°. At the camp the ants were very troublesome and annoying to-day; a strong wind was blowing from the eastward, and the camp was in a continual cloud of sand and dust. I would infinitely have preferred to be on horseback to-day, could I have indulged in my choice.

Tuesday, 4th November.—Thermometer last night 90°, this morning 60°. To-day was again windy and dusty, but not so hot as yesterday, the thermometer not rising above 90° in the shade, being no hotter than the temperature at eight o'clock last night.

Wednesday, 5th November.—Thermometer last night 74°, this morning 58°; a very windy, gusty night and cold morning. Thermometer in shade at three p.m. 85°. Late in the evening, while Jimmy and I were up at the water, shooting some pigeons, the three horses Mr. Tietkens had taken with him came up to drink, which of course informed me that he had returned. They looked fearfully hollow, and I could see at a glance that they could not possibly have had any water since they left. At the camp I was sorry to hear, that the journey had been quite unsuccessful; no water whatever was to be got anywhere, and what had drained at the Cob had now entirely ceased.

Thursday, 6th November.—Thermometer last night 66°, this morning 54°. A fine cool night, also a cool day. Thermometer 88°. Remained here again to-day. Where indeed was I to go!

Friday, 7th November.—Thermometer last night 74°; this morning 60°. To-day I started again for the creek, packing two horses with water in bags and one with rations, &c., intending to carry the water out to the creek, forty miles west from here; at that point I would water one horse, plant the remainder of the water in a tree, and follow the creek channel, to see what would become of it. It was after dinner when I started; taking Gibson and Jimmy Andrews, and leaving Mr. Tietkens alone in the camp. We travelled only fifteen miles, and camped in the scrub and triodia. The sky was cloudy, and threatened rain.

Saturday, 8th November.—Thermometer last night 74°, this morning 76°. The night was hot and cloudy, but no rain fell. At six miles from camp we crossed the ranges, that lay across our line of march, in a different place from where Mr. T. and I had previously done; we traced watercourses, and searched gullies, but every place was dry. At fifteen miles we turned the horses out for an hour, at which time the thermometer stood at 101° in the shade. Starting again we reached the creek below the junction in ten miles; it being too early to camp, I pushed on down the creek in five miles from where we struck it, to my great surprise—though the traveller in these regions should be surprised at nothing—we had completely run the creek out, as it simply ended amongst triodia, sandhills and scrubby mulga-flats. I was greatly disappointed at this turn of affairs, as I had thought that from its size this creek at least would have led me to some water, if not to the discovery of some new feature. Except where we first struck it the country had all been burnt, and we had to return to that spot to get grass to camp at. Water existed only in the bags which we carried with us. I gave the horse, which I intended riding to-morrow, a couple of buckets of water; the others of course had none.

Sunday, 9th November.—This was not destined to be a day of rest to my two men or myself, for move on we must; they to return to the shoeing-camp, while I went further on, for now that I had ascertained the termination of the creek, I resolved to push further west. The thermometer at night had been 74°, this morning 66°; the morning was cool and pleasant. Two of the horses were away, but the one I intended to ride was here. I gave him as much water as I could possibly spare (this was not as much as he would have drunk), and fixing another water-bag and leathern envelope, which held my future supply, containing about eight gallons, up in a tree, I started away, like an errant knight, on a sad adventure bound, though unattended by any esquire or shield-bearer, as was almost invariably the custom during the ages of chivalry. My course was west, and I continued on it over open triodia-sandhills, with occasional dots of scrub between, for twenty miles. The horizon to the west was bounded by open undulating rises of no elevation, but whether of sand or stone I could not determine. At this distance from the creek the sandhills mainly fell off, and the country was composed of ground thickly clothed with triodia, and covered all over with brown gravel. I gave my horse an hour's rest here, with the thermometer at 102° in the shade. There was no grass for him to eat, so he simply rested, not being possessed of digestive organs strong enough for triodia. Starting again and continuing my course, I entered upon country, where the triodia had all been but recently burnt. I travelled over miles and miles of burnt, stony, brown gravelly undulations; at every four or five miles I obtained a view of similar country beyond. At thirty-five miles from the creek the country both before and behind me was exactly alike, but here, on passing a rise that seemed a little more solid than most others, I noticed in a kind of little valley some signs of recent native encampments, and the feathers of birds strewn about. I of course rode towards them, and when close to them I saw, almost under my horse's feet, a most singular hole in the ground. Dismounting I discovered that it was another of those extraordinary cups, from whence the natives obtain their surface-water. This one was entirely filled up with boughs, and I had great difficulty in dragging them out, when I perceived that this orifice was of some depth and contained water; but upon reaching up a drop in my hand I found it was quite putrid—indeed while taking out the boughs my olfactory nerves gave me the same information—indeed the stench from it was revolting. I found it was choked up with rotten leaves, dead quadrupeds, birds and all sorts of imaginable filth. On poking a stick down into it, seething bubbles aerated through the putrid mass. And yet the natives had evidently been living upon this fluid for some time, and had only recently (if now) left it, as some of the fires were yet alight at their camps. (It is unknown, whether the aborigines employ charcoal or other antiseptic means, to purify such water.—F.v.M.) I had great difficulty in reaching down to bale any of this fluid into my canvas bucket. I found however, that my horse seemed anxious to drink, but one bucketful was all he could manage. There was but a small quantity of water in this reservoir—not more than five or six buckets at the most. There were fortunately a few hundred acres of silver-grass in the little valley near, and my horse commenced to feed with apparent relish; I therefore encamped here, though I anticipated at any moment to see a number of natives make their appearance; but as I had no intention of shifting my quarters, I did not fret much upon that account. I said to myself—

"Come one, come all, this rock shall fly
From its firm base as soon as I!"

However no enemies came, and I passed a tranquil night, with my horse feeding quietly close to where I lay.

Monday, 10th November.—I was up and away from this dismal burnt encampment before sunrise, not giving my horse any more of the disgusting fluid in the pit, as I felt certain it could do him no good. The night had been cool, and the morning felt especially so. At starting I continued my west course. In a mile or two I came to the top of one of those undulations, which at various distances bound the horizon. They are but swells a little higher than the rest of the country, the intervening spaces being hollows similar to those I have been travelling over, the country being all open with a few small belts of mulga in the hollows. There did not appear much likelihood of finding water in such a region as this; for where could it lodge? and how far this formation would extend was a question difficult to solve. What other feature lay beyond, on which water could be obtained, was equally difficult to guess. From the appearance of this region I was compelled to suppose, that it would continue unaltered for a very considerable distance. From this rise all I could see was a hollow for several miles; then another rise, which I reached in nine miles. Nearly all the country had been burnt, but what had escaped was simply triodia. The ground was still covered with gravel, with here and there small patches of scrub, the country in general being very good for travelling. From this rise I could see exactly the same view as that just traversed, that is to say, a hollow and rising ground beyond. I felt sure that it would be necessary to travel 150 miles at least before a likely spot could be found, in which to obtain water. How ardently I wished for a camel! for what is a horse, where waters do not exist except at great distances apart. I pushed on to the next rising ground, which was ten miles from the last, being nineteen from where I had camped. The view from here was precisely similar to former ones. I found that my horse had not travelled at all well this morning. He looked very hollow, when I started with him. I think the putrid water must have made him ill, for I had great difficulty in inducing him to go. I gave him half an hour's rest; it was not later than eleven o'clock when I did so. The day was very hot, the thermometer standing at 102° in the shade. There was a small belt of mulga with a little dry and withered grass amongst them, but the horse would eat none. To the south, about a mile from me, there was an apparently more solid rise, than where I was, so I walked to it in hopes of discovering another cup, but was quite unsuccessful. At this point I returned; for unless I was prepared to go a hundred or more miles, ten, fifteen or twenty in such a region would be no use. I was a long distance from my camp, it being distant forty miles from the creek, and I was now over fifty miles from that feature. I had a small supply of water certainly at the creek, but having been compelled to leave it hanging on a tree, it was at the mercy of the winds and waves, not to mention the possibility of its removal by natives. I had a great difficulty in getting my horse to travel back, the afternoon being exceedingly hot and oppressive. On reaching my old encampment I observed, by the removal of several boughs &c., that natives had been here in my absence, as they had put several of the boughs back into the hole again; the unfortunate wretches, I suppose, being compelled to use this frightful water. I had no doubt they were close to me now, though I could not see any of them or any fires; and I felt sure they were watching both me and my movements with lynx-like glances from their dark metallic eyes. My horse would not eat, but immediately hobbled off to where he knew water of some kind existed. I was afraid he would actually jump in, he seemed so eager for water. I drove him away once or twice during the night, for I could not sleep for thirst myself, as I had no water to drink. The horse however persisted in returning to the pit, so I thought to pass some of the time till morning by watering him. It was an exceedingly difficult operation, to get a bucket of water out of the abominable hole, as the bucket could not be dipped into it, nor could I reach the frightful fluid at all without hanging my head down with my legs stretched accross the top of the hole, while I baled the fœtid mixture into the bucket with one of my boots, as I had no other utensil; and what with the position I was in, and the horrible odour, which rose from the seething fluid, I was seized with most violent retchings. The horse gulped down the first half bucket with avidity, but after that he would only sip at it, and I was glad enough to find, that the one bucketful I baled out of the hole was quite sufficient for him. I don't think any consideration would have induced me to bale up another.

Tuesday, 11th November.—Having had but little sleep, I rode away at half-past three this morning back on my track to the creek, where my water-bag was hung. My horse looked wretched, and went very badly all the morning, insomuch that it was midday by the time I had travelled twenty miles, where, entering sandhill country, I was afraid he would refuse to go at all; but after an hour and a half's rest he appeared much better, and walked away almost briskly. I reached the creek at half-past five in the evening. Here we both had a good drink of good water, for nothing had disturbed the bag, which hung in the tree. Here I again encamped, the horse now feeding contentedly. The day had been very warm; the thermometer had stood at 102° at twelve o'clock in the shade. I had yet forty miles to go to return to the camp.

Wednesday, 12th November.—Giving my horse the remainder of the water from the bag this morning, I finally left this singular watercourse, where plenty of water existed in its upper portions, but it was either too bitter or too salt for use. Before departing however I named it Elder's Creek, in honour of the Hon. Thos. Elder, who has done so much for the exploration of the interior. The creek that joins from the eastward I named Hughes' Creek, after Captain Hughes, who has also, in connexion with Mr. Elder, contributed largely to the furtherance of western exploration. The ranges, where these creeks exist, I named the Colonel's Ranges, after my gallant old friend Colonel Warburton, the only one of three leaders of expeditions, who successfully carried out his mission, and to whom all honour is due as being the first white, who ever crossed Australia from east to west. There was not much water in the bag this morning, and the horse was standing close to it for some hours before daylight; therefore we made an early start for the depôt at the Shoeing-Camp, having forty miles to go. I reached it very late. Gibson and Jimmy had returned all right after leaving me, and everything was well at the camp except the water-supply, for Mr. Tietkens informed me, that the drainage was gradually ceasing, and that in a week probably there would be none. The day had been pleasant and cool, the thermometer not rising higher than 85° in the shade.

Thursday, 13th November.—The thermometer last night stood at 74°, this morning at 54°. It was a very cold and windy night. I took a day's rest to-day, to consider what next was to be done, as it was evident, if the water-supply failed, we must depart somewhere. It was not exactly a day's rest either, for Mr. Tietkens and I went up the gully and redug our tank, hoping against hope thus to increase the water-supply. I could of course fall back on waters, found more easterly. The horizon in that direction always looked so inviting, ranges piled on ranges, stretching as far as the vision could go; but then I had come from there, and my mission was to go west. I had now good reasons to conclude, that I could not do so from this point; it was evident a more northerly or more southerly one must be found, even if I had to retreat some distance eastward to discover another line.

Friday, 14th November.—Thermometer last night 66°, this morning 52°; a fine cold night. This morning I took Gibson with me, and we went away to the southward, as there was a long abrupt-faced line of range, though of no great elevation, running along the horizon in that direction. I was most anxious to find more water, if I possibly could, as our supply at the camp was now failing fast, and I could not depend upon it even for twelve hours. There was a kind of notch or pass in the range lying nearly S.S.W.; so we travelled towards it, and reached it in fifteen miles over sandhills and through scrubs as usual. I had hoped this was a kind of table-land extending southward, but found it consisted only of a single ridge, having an abrupt face to the north; it was covered with cypress-pines, and where we struck it there was a small creek-channel, but it was dry. The view from the summit to the south was cheerless enough, as though there were a few isolated ridges within a few miles, yet beyond them the country appeared all scrubs, with the reddish tops of burnt sandhills amongst them; a line of bloodwood or other eucalypts ran along east and westerly some four or five miles off, which appeared to come out of these hills further to the west. We therefore descended to the southern foot of this range and followed along it to the westward, and in five miles we came upon the creek seen from the hill. There was a thick fringe of teatree-bushes along the channel; on reaching which however we found it perfectly dry; the whitish marks of the wash of water gave me some hopes of yet discovering some further up the channel. Another line of hills on the opposite side of the creek here formed a kind of glen; it was a very pretty spot, and it seemed most inviting for water, but having followed it up to its head, we were quite unsuccessful in finding any, nor was any to be obtained by digging. I next followed it down for some distance, and by the time we had come opposite to the gap through which we came, we had travelled twenty-five miles; so we turned the horses out for an hour. The day was comparatively speaking cool, the thermometer not rising higher than 90° in the shade. Starting again we followed further down the creek until it was quite exhausted amongst the sandhills, which environed it, and we camped in the dry channel, having travelled forty miles. The country about here was thickly covered with burrs, and it was only on the bare sand of the creek-channel that we could escape them. They may be said to infest this part of the country.

Saturday, 15th November.—The night was exceedingly cool, if not cold, the thermometer at night standing at 68°, by morning at 40°. I was greatly disappointed at finding no water in this creek; the country looked well near it, and the teatree grew vigorously along its banks. As every day was telling upon the water at the camp, being unsuccessful here, I hastened back. We had thirty miles to go, and reached it early in the afternoon. Mr. Tietkens informed me the water would scarcely suffice, to give our three horses a drink; that he had been hard at work during our absence deepening the tank; but as that did not increase the supply, we should be compelled to retreat to the eastward. I called the range, to which Gibson and I had been, the Rampart-Range; but it is probable Mr. Gosse named it, as he must have seen it before I did. We saw a great number of pigeons about it, and no doubt they are aware of the existence of rocky basins from which they obtain their water, but they did not convey any information to me upon the subject. The day was much warmer than yesterday, the thermometer standing at 98° in the shade.

Sunday, 16th November.—Thermometer last night 64°, this morning 66°. I had determined to retreat this morning from this camp, where we had been compulsory residents so long; but the horses all running loose, and the water being so low, only one animal could water at a time, and those that got water in the night had rambled away, whilst most of the others were up the gully waiting for their turn to drink. We could not get them altogether to-day, so had to remain. The thermometer rising to 102° in the shade.

Monday, 17th November.—Thermometer last night 74°, this morning

58°. A cold windy morning. We had work to-day in getting all the horses together, so much so, that it was five o'clock p.m. before we could get away. The water had entirely ceased to drain into the tank, and some of the horses had gone many miles away in search of more, but found none. I did not like the idea of retreating the way I came, though I could scarcely hope to find any water to remain nearer than that at Gosse's Depôt Range, and as the hills, on which we were now, trended north-easterly, I decided to run them up, in the first place to look for water, and secondly to discover how far they ran in that direction. It appeared to me, I should be compelled to go formally into depôt-quarters for a time in hopes of some rain falling, when I could start out again to the west. To the east I might go at any time, as in that direction I had left some permanent waters behind me. I was under the impression, that I might from a more easterly point discover other lines of ranges in a north or north-westerly direction, which might extend perhaps across the entire continent; and those in which I was now, might perhaps lead me to them; therefore, as I said before, I determined to follow up their north-east trend. There were several high hills in the ranges now to be traversed, and I was not without hopes of discovering some water, which would enable me to form a depôt amongst them. As it was so late, when we left the Shoeing-Camp, we accomplished five and a-half miles by night and encamped in a rather scrubby valley and without water. The day, though warm, was cloudy; I was too busy to notice the indication of the thermometer, but it could not have been higher than 90° in the shade.

Tuesday, 18th November.—Thermometer last night 70°, this morning 58°. The morning was cloudy. Having travelled in a north-easterly direction fourteen miles through scrubby valleys and over scrubby hills and many triodia-sandhills, we crossed several small watercourses and gullies, which were all perfectly dry. In searching up one small creek I thought to have discovered a prize of sufficient value, to give all my horses a drink, of course by digging out a tank; but while I was in the hills searching, the party had gone on two or three miles, at which distance I overtook them, and immediately dispatched Gibson back with the shovel to open out the place. When we arrived he gave a very gloomy account of the water supply; but as there was some, and it drained in slowly, we unpacked the horses; but unfortunately the supply was so wretchedly limited, that we had to take it in turns to watch the tank and give each horse only a bucketful as it drained in. I was much vexed at having brought them back to such a miserable spot; only half of them got their bucketful by twelve o'clock at night. The day had been cool, the thermometer not going higher than 90° in the shade.

Wednesday, 19th November.—Thermometer last night 70°, this morning 68°. The horses were all round about the tank this morning; those who had had their bucketful seeming more greedy than those who had none. It was half-past ten o'clock by the time that the last horse got his bucketful, and then we left this miserable little glen, and proceeded on our yesterday's track again to where I had overtaken the party. Travelling still on the same course as yesterday we traversed scrubby hills, triodia-sandhills and scrubby valleys, searching every place, which had the appearance of ever having had water at it, but found none. I had been steering mainly for a big hill, since I left the Shoeing-Camp, and now at thirty miles further we reached it; there was a small creek-channel coming down from it, but it was utterly destitute of water. From its summit I saw that these hills went no further in a northerly direction. A low ridge lay out to west, and only to the east and south-eastward could ranges be seen. I was therefore compelled to go where ranges lay, for in this region it is only in ranges, where water can be found. Mr. Gosse's Depôt Range, at least that portion of it, from whence I had issued, lay 30° south of east, some forty-seven or forty-eight miles away, and as that was the nearest place at which I knew water to exist, I departed for it. A detached jagged and inviting range lay also a little to the east of north-east; it had all the appearance of having waters about it, as it was of similar outline and formation to the hills at Gosse's Depôt Range. I called it Jamieson's Range, after William Jamieson, Esq., of Mildura, on the Murray, an old and most kind friend of mine, and a subscriber to the fund for this expedition. It being late in the day, when we left the big hill, we only got twelve miles from it by night, having travelled twenty-four miles; a great deal of time was lost in useless searches for that element, which did not exist here. The day was very warm, the thermometer rising to 104° at ten o'clock in the morning. From mid-day the sky became greatly overcast, and the atmosphere was close and sultry. The last twelve miles from the big hill was through more open country, being level sandy ground, with porcupine-grass. When we camped we had thunder and lightning, and about half-a-dozen heatdrops fell. The horses were very thirsty, and our own supply consisted of about a quart of water between four of us.

Thursday, 20th November.—Last night was close and sultry, the thermometer standing at 80° at night and 76° in the morning. We made an early start and proceeded on towards Mr. Gosse's Depôt Range on our yesterday's course. At fifteen miles we had a shower of rain, which lasted perhaps a few seconds, with thunder and lightning; we were near a red bare rock, and had it lasted with the same force for only one minute, we could have given our horses a drink upon the spot; but as it was, we got none, though the horses ran about the rocks, licking the stones with their tongues. Pushing on, we reached our old encampment late at night. We had formerly obtained water here by digging, now we had to go further up the creek to get any. I intended to tie the horses up until we dug a tank, but while I went up the creek to find a good place, the others unpacked most of the horses, and when I returned I met them in the channel hobbling along up the creek, and if they went far enough, I knew they would get plenty of water; but now the sky was as black as pitch; it thundered and lightened, and there was every appearance of a fall of rain, so the remainder of the horses were also let go, hoping they would go themselves up to the surface water—for, with the exception of one bucketful each, this was their fourth night without any. A kind of light mist or heavy dew fell for an hour or two, but it was so light and the temperature so hot, that we all lay without a rag on and the rain all passed away.

Friday, 21st November.—All the night was close, cloudy and sultry; the thermometer at night standing at 90°, in the morning at 76°. Mr. Tietkens and I were up at first dawn this morning, and taking the shovel with us up the creek, to where we heard the horse-bells, we found most of the horses together; about twelve of them were lying down in the bed of the creek with limbs stretched out as if dead; on coming up to them we were glad to find they were all alive, but some of them could not get up; some that were standing up were working away with their feet stamping out the sand, trying to dig out little tanks, and one old stager had actually reached the water in his tank, so we drove him away, and dug a proper tank, where he had shown us the water; the supply was pretty good, and we got all the horses watered by nine o'clock—it being four a.m. when we commenced digging; this little exercise gave us an excellent appetite for our breakfast. At one o'clock the thermometer stood at 102° in the shade, Gibson having built a bough gunyah, under which we sat. I anticipated being compelled to remain here some time. In the afternoon the sky became cloudy and overcast, and at six o'clock in the evening rain actually began to fall, and it came down heavily for about a quarter of an hour; it continued to drip for a couple of hours, but after the first quarter of an hour, it might as well have stayed away; it however cooled the atmosphere, and we had heavy thunder and lightning for a long time after the rain ceased. The thermometer at nine p.m. fell to 48°, and we felt it quite cold; at this time the sky again cleared. A slight trickle of surface-water came down the creek channel. I was greatly in hopes more rain had fallen to the west, as from that direction the rain came. The thought of being shut up here for months in depôt was most agonising; the only hope to console myself was, that more rain might have fallen to the west; at all events I resolved to push out there again and see.

Saturday, 22nd November.—Thermometer last night 48°, this morning 58°. The rain had only fallen lightly, but the sky was overcast again, and though the water had trickled down along the creek-bed soon after the rain, there was no surface-water this morning, but the tanks we had dug were now quite full. The horses were all loose and feeding at their pleasure. The day was most agreeable, the thermometer not rising more than 80° in the shade.

Sunday, 23rd November.—Thermometer last night 64°, this morning 54°. Being Sunday, we did not remove the camp, and spent the day in peace and quietness. I determined to leave here again to-morrow, and revisit both the Shoeing-Camp and Elder's Creek country, as I believed more rain had visited that quarter, than we had here, and it appeared to have been raining over to the westward the day before we had any here.

Monday, 24th November.—Thermometer last night 66°, this morning 64°. We had thunder, and a few minutes' sprinkling rain again last night, but of no volume; this morning was fine and clear, but towards mid-day it became warm and sultry. Made another departure from this range *en route* to the Shoeing-Camp; but it being late by the time we got away, we only accomplished sixteen miles upon the bearing we had last arrived here upon, that is to say, 30° to the N. of W. At this spot, as I mentioned previously, there were some bare rocks and a dry native well, and where we had rain for a moment only. Upon reaching this rock, I now found the rain had left sufficient water and more for all my horses, as the native well was full and there was water also lying upon the rocks. This little spot now looked quite a pretty and pleasant place; very different to its original appearance, when we found no water at it, so wonderful a difference does the presence of that all-important fluid create. Soon after unpacking the horses, another thunderstorm came up, and the rain soon refilled the basin, which the horses had considerably reduced; the rain continued nearly five minutes; after this the evening was dry and the sky clear. I called this little spot the Lightning Rock (not that there is any similitude between it and the so-called rock at Port Phillip Heads), because when we passed here previously and on this visit, the lightning played so vividly around us.

Tuesday, 25th November.—Thermometer last night 66°, this morning 54°. A fine cool and dry night; but just as we were starting away from here more thunders and lightnings occurred, also the usual duration of rainfall, namely about five minutes. This storm apparently came from the west, and went by us northward. From here I steered N. 80° W., to strike my former tracks, where we had obtained the one bucketful of water for each of the horses; and after travelling sixteen miles, we found some water actually lying upon the surface of the ground, a most extraordinary circumstance. The ground had been rather soft for some miles, and I had hopes that the country to the west had been well visited with rains. The whole country traversed to-day consisted simply of dense scrubs; and to enable the other members of the party to bring the horses along with certainty, I always rode in advance, having a horse-bell fixed on to my stirrup, so that those behind, though they cannot see, yet can hear, which way to come; continually working this bell with my foot has almost deprived me of the faculty of hearing. The continual passage of the horses through these abomiable scrubs has worn out more canvas-bags, than ever entered into my calculation. Every night after travelling, some if not all the bags are sure to be ripped—causing the loss frequently of flour and various small articles that get jerked out, when these rippings take place. This has gone on to such an extent, that every ounce of twine I had with me has been used up, and I was at my wits' end to discover some method, by which further damage might be prevented or more sewing material obtained, and the only device I could hit on was the unravelling of canvas, of which fortunately I had some little supply. Ourselves and clothes, as well as pack-bags also, got continually torn. Any one in future traversing these regions must be equipped in leather; there must be leather shirts, leather trousers, leather hats, leather heads and leather hearts; for nothing else can stand the frightful tearing wear of such a region as this.

Wednesday, 26th November.—Thermometer last night 70°, this morning 56°. A considerable dew-fall occurred last night. We continued on our yesterday's course for the one-bucket-place, but in consequence of deviations from the right line to avoid hills, we struck our former tracks in sixteen miles, about two and a-half miles south of that spot, at another small creek I had also formerly inspected; and I made sure of finding water at it, as it had a few little clay-holes in it; but on reaching it I was surprised to find, that not a drop of rain had fallen here, and I began feel alarm lest the Shoeing-Camp should also have been unvisited. We turned now on our old tracks south-westward for the Shoeing-Camp. One of the horses had been unwell, and concealed himself in the scrubs, and it was some time before he was missed; so we had to encamp in some open triodia-ground without water, while Mr. Tietkens and Gibson went back to look for the missing horse; they brought him about dusk. There was no grass here for the horses to eat. The day had been agreeable and cool.

Thursday, 27th November.—Thermometer last night 64°; this morning it was the same. The night was cool, but no dew fell. We arrived at the old camp by twelve o'clock, and there had evidently been a little rain. Mr. Tietkens went up to see how the tank looked, and it was with great difficulty that we could keep the horses from rushing up the gully with their packs on after him; but having put up a small yard, when here previously, we drove them into it. The horse that concealed himself yesterday knocked up this morning, and we had to put his pack on other horses. Gibson stayed behind to bring him on, but did not arrive for four hours after us, though we only left him three miles behind. Mr. Tietkens returned from the tank and informed me, that though it had been filled and was now running over, yet he thought the drainage would scarcely last a week, as neither above or below it was there any signs of water. The thermometer went up to 102° in the shade to-day. There was not sufficient water in the tank for all the horses, therefore some of them had to wait, until sufficient drained in, which it did but slowly. I was exceedingly grieved to find, that so little water appeared after the rain; I had thought, that a greater quantity had fallen to the west, than where we had been, as all the storms appeared to come from the western quarter. This camp had now become simply a moving mass of ants; they were bad enough when we left, but now they were frightful; they swarmed over everything, and bit us to the verge of madness. It was eleven days since we left this place, and now having returned it seems highly probable that I shall soon be compelled to retreat again.

Friday, 28th November.—Thermometer last night 80°, this morning 68°. The early part of the night was very warm; the ants were unbearable, at least they were to Mr. Tietkens and myself; Gibson and Jimmy did not appear to lose any sleep on their account, though I could not lie down for more than a moment or two at a time in any one place. Towards morning it became a little cooler, and the ants were not quite so persistent. The day was again warm; the thermometer stood at 102° in the shade at three p.m. With the aid of a quart-pot and tin-dish I managed to get some sort of a bath, but this is a luxury the traveller in these regions must learn in a great measure to do without; but I could no longer restrain myself, for my garments and person were so perfumed with smashed ants, that I could almost have believed I had been bathing in a vinegar-cask. We remained in this camp to-day.

Saturday, 29th November.—Thermometer last night 80°, this morning 76°. A very warm night. I did not think it advisable to start away with all the horses to the west without knowing how the rains had fallen, so I dispatched Mr. Tietkens and Jimmy to take a tour round to all the places we had formerly visited out west. At twenty-five miles was an almost bare rocky hill, which I called *par excellence* The Cups, as having so many of those singular little stone indentures upon its surface, and which I first discovered on the 19th October. If no water was there, I directed Mr. Tietkens then only to visit Elder's Creek and return, as, if there was none at the Cups, there would be very little likelihood of any in any other place. Gibson and I had a most miserable day in the camp; the ants were dreadful, the hot winds kept blowing clouds of sandy dust all through and over the camp. The thermometer rose to 102° in the shade. I was hard at work unravelling canvas for twine, while Gibson was repairing packbags with it.

Sunday, 30th November.—Thermometer last night 86°, this morning 64°. The early part of the night was hot, and some few mosquitoes troubled us; but their stings and bites were delightful pleasures compared to the agonies, inflicted on us by the myriads of small black ants; towards morning it was pretty cool, and most of the ants retired. A hot wind was blowing again to-day, and we were again smothered in dust at the camp. There was so much sewing required amongst the packbags, that although it was Sunday, Gibson and I were compelled to work at them to get them into something like order and usefulness. Thermometer in the shade to-day 100°. The water-supply in the gully is again failing, the tank not having filled during the night.

Monday, 1st December.—Thermometer last night 86°, this morning 74°. Another hot night. To-day about one p.m., Mr. Tietkens and Jimmy returned from the west, and reported to me, that the whole country in that direction had been entirely unvisited by rain, with the exception of the Cups, and there they obtained out of several dozens barely sufficient water to satisfy their three horses. Elder's Creek, the tank at the Dead-Cob, the Colonel's Range, Hughes' Creek and the range lying between here and the Colonel's Range (the way Mr. Tietkens and Jimmy returned) were all alike perfectly dry, not a drop of moisture having fallen in that direction. To-day was exceedingly hot; the thermomometer indicating 106° in the shade. Our water-supply still gradually decreasing. It is evident, I must again retreat from here; for in two days every drop will be gone. This is certainly not a delightful position to hold; indeed it is one of the most horrible of encampments. The small water-supply is distant about a mile from the camp, and we have to carry it down in kegs on a horse to the camp, and often when we go for it, we find the horses have just emptied it. We are eaten alive with flies, ants and mosquitoes; so our existence here cannot be deemed a happy one. Whatever could have obfuscated the brains of Moses, when he omitted to inflict Pharoah with such exquisite torturers as ants, I cannot imagine. For I fancy I could wallow in blissful repose in darkness amongst cool and watery frogs, but ants, oh! they are frightful!

Tuesday, 2nd December.—Thermometer last night 96°, this morning 86°. A very hot night and cloudy morning. The water-supply alarmingly small this morning. Like Othello I am "perplexed in the extreme," for the sky threatens rain every day, I don't like to go, and I can't stay; a hot wind again blowing; thermometer 104° in the shade. Mr. Tietkens and I went out over the hills, to see if we could find any other water anywhere in the range, at length we found an old native well in a small creek a little S.E. of our camp—we thought by digging and removing a quantity of stones we might get sufficient water for a day or two at least; we had more thunder and lightning to-day and a few drops of rain.

Wednesday, 3rd December.—Thermometer last night 82°, this morning 68°; a much cooler night than the previous one. Mr. Tietkens and I went to another place to-day and dug a tank, and worked for some hours removing a vast quantity of stones and soil, but so little water coming in, we at length abandoned it. The water at the camp is all gone and the horses had been standing over the tank all night, some few had a partial drink and had gone away to feed; otherwise I would have packed up and gone away to the place Mr. Tietkens and I found yesterday, but it being now nearly night when we returned from fruitlessly digging, I was obliged to wait until morning. The day had not been excessively warm, the thermometer not rising higher than 100° in the shade.

Thursday, 4th December.—Thermometer last night 88°, this morning 70°; the night was again hot. Having got all the horses, we made a second retreat from this deceitful spot; it was about 2½ miles to where we had seen the native well, and we soon got there. Mr. Tietkens having gone ahead with the shovel to open out a tank, on arriving we found but a poor supply at the place where he had dug, but we opened out several places, and towards evening all the horses had been satisfied. We shifted some monstrous blocks of stone with levers, and as they left deep holes I expected a better supply by morning. There was a hot tropical sultry feeling in the atmosphere to-day, though the thermometer never rose above 100° in the shade: in the afternoon we had thunder and lightning; rain also fell for a few moments towards night, when the atmosphere became less oppressive.

Friday, 5th December.—Thermometer last night 76°, this morning 68°. The sky was again very cloudy this morning and threatened rain. I was glad to perceive, that the supply of water in all the tanks we had made, was as good as could be expected, and they gave a full drink for all the horses, though I can scarcely hope the supply will keep up long, for this is such an extraordinary region, one never can tell when the drainage may cease; your tank may be full, and you may congratulate yourself upon the supply of a morning, but when your horses empty it, it may never become replenished. My present camp is in a most wretched place, all scrub, stones and porcupine-grass. The thermometer rose only to 100°, but it was close and sultry; we had a breeze from the north at night; there were also terrific lightnings in that quarter for several hours, but we heard no thunder.

Saturday, 6th December.—Thermometer last night 86°; this morning those figures were transposed. I was very doubtful of the continuance of the drainage here, and wished to find some other place, where water could be obtained, without going back again so far as Mr. Gosse's Depôt-Range. Mr. Tietkens and I therefore went out eastward, towards the hills I had formerly called Gosse's Pile, because when I passed through them before in October, I had found a small creek with a good supply of drainage-water in the sand; and as there was a good open piece of country near it, I thought, if the water was still there, that it would be a good place to remove to, when our present supply failed. To reach it we traversed a region of scrub, and found the creek-bed was perfectly dry. We searched then up amongst the hills, but not a drop of water could be found. The thermometer stood at 100° in the shade. The sky at three p.m. became overcast, and we had thunder and lightning with a few drops of rain. After our unsuccessful search in the hills we returned down the creek, and camped near our original outgoing tank, now however without water.

Sunday, 7th December.—Thermometer last night 80°, this morning 76°. There were some ridges to the south, and the creek we were upon ran down towards them; we therefore followed it down in that direction. The creek however ended before we reached the ridges. There were a few cypress-pines upon them. There were also many stony gullies, consisting of numerous loose blocks of stone, but all perfectly dry and waterless. After fruitlessly searching among them for water we turned homewards again. The day was warmer than the last two or three had been; the thermometer rose to 106° in the shade. At three p.m. we had thunder and lightning and a few drops of rain as usual. On reaching the camp at night I received the intelligence, that the water-supply was failing rapidly, becoming fine by degrees and beautifully less. At night there were vivid lightnings to the south for several hours, but too distant to permit the sound of thunder to be heard.

Monday, 8th December.—Thermometer last night 86°, this morning 82°. A very warm night followed by a very hot day, the thermometer standing at 112° in the shade. We had clouds, thunder and lightning, but no rain. The water is failing so rapidly, that to-morrow I must again retreat, as here I can no longer remain. The horses seemed never satisfied. They came two and some of them three times during the day for water, and would have remained round the tanks altogether, had we not been compelled to drive them away.

Tuesday, 9th December.—Thermometer last night 80°, this morning 78°; another warm and uncomfortable night, but here fortunately we were not troubled with ants. This morning no water had drained into the tanks, so the horses could get none. They all came up of their own accord this morning except three, which Gibson went to find, and we got away by ten o'clock. The day was again hot, the thermometer standing at 110° in the shade. Finally leaving these miserable hills I steered away on a course of 10° north of east, intending to return to Mr. Gosse's Depôt-Range, the only place in this region where I knew we could obtain water, *via* the Lightning-Rock. When we formerly left that place the native well had been filled up, and I hoped to get water there now. It was forty-five miles from here. We traversed a country nearly all scrub, passing hills, and searching several water-channels and gullies for water, but found none. We travelled twenty-one miles, and camped on a small dry water-channel, not to say a creek. I had felt very unwell for the last four or five days, and to-day I was almost too ill, to sit on my horse. I had fever and rheumatic pains all over. The country being all scrub, I was compelled to ride with a bell as usual on my stirrup—jingle-jangle, jingle-jangle all the day. What with the heat and the fever I was in, and the din of the bell, I really thought it no sin to wish myself out of this world, and into a better, cooler and less noisy one, where—

> "To heavenly harps the Angelic choir,
> Circling the Throne of the Eternal King"

should

> "With hallowed lips and holy fire,
> Rejoice their hymns of praise to sing,"

Which revived in my mind vague opinions with regard to our Christian notions of Heaven. It is doubtful if the thought should be so pleasing as the opinions of Plato and other philosophers, and which Addison has rendered to us thus:—

> "Eternity, thou pleasing dreadful thought,
> Through what variety of untried being—
> Through what new scenes and changes must we pass?
> The wide, the unbounded prospect lies before me"—&c.

But I am trenching upon debatable ground; I have no desire to enter upon an argument on the subject; it is doubtless better to believe the tenets, taught to us in our childhood, than to seek at maturer age to unravel a mystery, which it is self-evident the Great Creator never intended that man in this state of existence should become acquaiuted with. However I'll say no more on such a subject; it is quite foreign to the matter of my travels, and does not ease my fever in any way, in fact it rather augments it.

Wednesday, 10th December.—Thermometer last night 92°, this morning 76°; the early part of the night was hot, but it cooled considerably towards morning. I was terribly ill this morning, and it was agony to have to rise, but go I must. The day was again hot, the thermometer standing at 110° in the shade. We reached the Lightning-Rock by three p.m. in twenty-four miles, on the same bearing as yesterday; the water was all gone from the surface of the native well, but by digging a small quantity was obtained. I was too ill to dig, Mr. Tictkens and Gibson doing the work; there were a number of native fig-trees growing upon the rocks, and while Gibson was at work, Mr. Tietkens went to see, if he could get any ripe fruit; and it was most fortunate he did so, as on the side of the hills, opposite to where we were, he noticed another large bare rock, on which he also found a fine large rock-tank, with a good supply of water in it. He did not get any figs, but the discovery of the tank was far more preferable; we soon drove all the horses over to it, and they were all enabled to drink their fill. I was never more delighted at the discovery of water in my life, for there was sufficient to allow me to remain here a day or two, the thought of riding again to-morrow was killing to me; indeed I had made up my mind, to remain behind, while the rest of the party and the horses went on to the water; but this discovery enabled us all to remain.

Thursday ,11th December.—Thermometer last night 80°, this morning 62°. To-day the camp was removed to the new tank, about a mile. I could scarcely move hand or foot. The day was very warm, the thermometer standing at 108° in the shade.

Friday, 12th December.—Thermometer last night 86°, this morning 63°. To-day we remained at this rock-tank. It was most fortunate that there was sufficient water for us, though the horses had now reduced it most considerably, and we shall not leave a drop behind us when we depart; this basin would be of some size, were it cleaned out. We could not tell what depth it is, as it is now almost entirely filled up with the debris of ages. Its shape is elliptical, and it measures thirty feet long by fifteen feet broad, its sides being even more abrupt than perpendicular, and the horses could not get to the water without jumping in at one place; there was only about two and a-half feet of water in this basin, the rest being soil. If I possessed a wheelbarrow or a cart, I would have done something to it; but without such appliances nothing could be done. The day was much cooler, than the preceding ones; the thermometer only rose to 100° in the shade.

Saturday, 13th December.—Thermometer last night 34°, this morning 70°. We remained at this tank again to-day. I called it Tietkens' Tank. The thermometer rose to 104°. The water I am sorry to say is nearly done. I am better to-day. To-morrow we must again depart, as by that time all the water will be gone.

Sunday, 14th December.—Thermometer last night 86°, this morning 74°. The water is gone, the tank dry, and the horses are away to the east. It was past 3 p.m. when they were brought back. Unfortunately Gibson's little dog Toby followed him out to-day, and never returned. I sent Gibson back to the Rock (after we started) to await his return, while we went on, but at night Gibson arrived without Toby; no one knows what became of the poor little animal. I told him he could have any horses he liked to go back for him to-morrow; but Gibson was taken ill to-night, much the same as I had been, therefore poor Toby was never recovered. We have still one little dog of mine brought by me from Adelaide, of the same kind as Toby, that is to say, the small black and tan English terrier, though I am sorry to say, he is decidedly *not* of the breed of that Billy indeed, who used to kill rats for a bet. I forget how many one morning he ate, but you'll find it in sporting books yet. There was no water near the surface at our old tanks, when we arrived, though I knew there was plenty up the creek, and should have gone up to it, but the unpacking was commenced at the old camp, while I was behind. It was very late when we reached the place, and as I was too ill to dig, and Gibson still behind, we turned the horses up the creek, hoping they would go to the water. We had to re-dig one of the tanks, to get water for our own use. For the second time I have been compelled to retreat to this range. Shall I ever get away from it westward? is the question I am always asking myself. Evidently it requires rain, as I now know the country for 150 miles from here, and I am morally sure that there is not a sufficient quantity in all that distance now to water a single horse; and how far it may be beyond 150 miles to the next water, is more than I have been able to determine. The day was again very hot, the thermometer standing at 108° in the shade before we left the Lightning-Rock.

Monday, 15th December.—Thermometer last night 88°, this morning 78°. I was a little better to day, but Gibson was now very ill and could not move. The camp was therefore not shifted. Mr. T. and I went to look for the horses. We got three or four close handy, but the others had gone in various directions, mostly back along Mr. Gosse's dray-track, instead of going up the creek to water. By afternoon we got them all and took them up to the surface-water in a narrow glen, through which the creek runs; we took the horses up in two lots. When we arrived with the first mob there was plenty of water running merrily along the creek-channel, and several nice ponds full, but when we brought the second mob, an hour and a-half afterwards, the water was gone, the stream had ceased to flow, and the nice ponds just mentioned were all but empty and dry; this completely staggered me, so suddenly to find the drainage cease. The day was again warm, the thermometer standing at 110° when we returned to camp. I was in a perfect bewilderment at the thought of the water having so suddenly disappeared, and I was wondering where I should have to retreat to next, as it appeared there would be literally no water here at all in a day or two. On reaching the camp I was much exhausted with my morning's walk, and lay down in the shade of our bough-gunyah, which Gibson had erected the last time we were here. I had scarcely settled myself upon my rug, when suddenly a most pronounced shock of earthquake occurred; the volcanic wave, which caused a sound like thunder, passing along from west to east right under us, shaking the ground and the gunyah so violently as to make me jump up. As the wave passed on we heard up in the glen, to the east of us, great concussions, and the sounds of smashing and falling rocks hurled from their native eminences, rumbling and crashing into the glen below. The atmosphere was very still to-day and the sky clear, except to the west, where a few clouds were gathered.

Tuesday, 16th December.—Thermometer last night at 92°, this morning 82°; the night was exceedingly hot and close, the morning cloudy; a few sprinkling drops of rain fell. Gibson is still very ill; so I shall not remove the camp to-day. I was of course in a great state of anxiety about the water-supply; Mr. Tietkens and I therefore went after the horses, and getting them all together we took them to the place where we had watered them yesterday, when I was enchanted to behold the stream again in full flow, and the sheets of surface-water as large and fine as when we first saw them yesterday. I was rather puzzled at this singular circumstance, and concluded that the earthquake had shaken the foundations of the hills and thus forced the water up; at all events, from whatever cause it proceeded, I was exceedingly glad to see it (I, of course, had heard of intermittent springs, but I could not consider this as one of that class). To-day was much cooler than yesterday, there was an east wind blowing, the thermometer did not rise beyond 100° in the shade. At 3 p.m. (being at the same time of day) we had another shock of earthquake similar to that of yesterday, only that the volcanic wave passed a little more northerly, and the sounds of breaking and falling rocks came from over some hills north-east of us.

Wednesday, 17th December.—Thermometer last night 82°, this morning 76°. The night was cloudy, and we had thunder and lightning towards morning. Gibson being better to-day I removed the camp up to the glen, where the surface-water existed. We pitched our encampment upon a small piece of rising ground, and where there was a fine little pool in the creek-bed, partly formed of rocks, over which the murmuring streamlet flowed, forming a most agreeable little bath. The day was cool and cloudy, the thermometer not rising higher than 100° in the shade. The glen here is almost entirely filled up with teatree, and we had to cut quantities of it away, to approach the water easily. This is also the only kind of timber here for firewood. Many trees are of some size, being seven and eight inches through, but mostly very crooked and gnarled. The green wood appears to burn equally as well as dead, and forms a good ash for baking dampers. We had our usual shock of earthquake again to-day, and at the usual time.

Thursday, 18th December.—Thermometer last night 82°; this morning 74°. A windy and cloudy night; the day cool; thermometer 100° in shade. Three p.m.: earthquake, quivering hills, broken and toppling rocks, and scared and agitated rock-wallabies. We erected a fine large bough-house to-day, as they are so much cooler than tents. We also cleared several patches of rich brown soil, and made little gardens, planting all sorts of garden and other seeds. I have now discovered, that towards afternoon, when the heat is greatest, the flow of water ceases in this creek daily, but at night, during the morning hours, and up to about midday, the little stream flows purling along over stones and through sand as merrily as one can wish. This encampment, which I have named "Fort Mueller," cannot be said to be a very pretty spot, for it is so confined by the hills on both sides of the glen, that there is scarcely room to turn round in it. The hills are formed of huge masses of red and black blocks of bare stone, being granite, coated with iron. We have had also to clear the ground round our fort of the triodia, which originally grew here. The sides of the hills are also thickly clothed with this dreadful grass. There are some fine open flats and valleys round this range, where the horses feed; indeed, it is an excellent piece of country, except immediately near the water or on the hills, and will probably not escape the notice of the first settlers in this region. Mr. Gosse's camp is in the same range, and about four and a half miles away to the westward. It was his last depôt-camp, and numbered 14; and it is not at all surprising, that he returned, having searched the country well for some distance beyond this point; and the view of any country west from here is of such a gloomy nature, that all I can say about it is, that it is exceedingly *inviting to return.* From here I pushed beyond the extreme points of all hills, watercourses or scrubs, and entered, as I have previously described, an open, undulating triodia-desert, which ends I cannot judge where. I am now compelled to rest awhile in depôt, hoping, as we are so constantly threatened, that rains will fall and enable me, to push out westward again. I only ask for water at Elder's Creek, a hundred miles west from here. The sky has been greatly overcast all the morning, especially to the northward, and I thought, by taking a tour out in that direction, I might find some places, where rain had fallen; so Mr. Tietkens and I prepared to start away again to-morrow.

Saturday, 20th December.—Thermometer last night 74°, this morning 66°. Though the sky was cloudy and overcast nearly all night, the morning was clear and cool. Mr. Tietkens and I got away rather late, and only travelled eighteen miles, passing through, or rather over, the western end of another line of hills lying N.E. from here, whose extreme western point I named Mount Scott, after Andrew Scott, Esq., of Jolimont, Melbourne, a subscriber to the fund for my expedition. We found a small creek, but no water. The country appeared to have been quite unvisited by any rains. The day was cool, the thermometer rising only to 90°. We had to encamp without water for the horses. We carried a small quantity in a keg for ourselves. The country between these two ranges was not nearly so scrubby as that lying further west. It consisted chiefly of red sandhills, which the natives had recently burnt, and the country was free from spinifex, on that account making travelling the more easy. Our course to-day was north 40° east.

Sunday, 21st December.—Thermometer last night 74°, this morning 62°. I felt the temperature exceedingly cold last night, as I had only brought one small rug with me from Fort Mueller. In the morning we steered away for a dark-looking, bluff-ending hill, which bore north 20° east. We reached it in twenty-two miles; but before arriving at it we searched amongst a lot of pine-clad hills for water, but found none, nor was there any to be got at the Bluff-Hill. We rested our horses here for an hour, during which time we ascended the hill. From it I discovered, that to the north and for 30° east of that point, and also 10° to the west of it, lay a number of ranges at a very considerable distance from us. The nearest which I considered the most northerly to be I could not estimate at less than sixty miles. They appeared of some length, but were not sufficiently raised above the horizon of scrubs (which occupied the intervening regions, and which also rose in undulations to some elevation) to enable me, to form an opinion with regard to their altitude. Those east of north seemed further away and higher, being bolder and more pointed in outline. None of them, I should have remarked, could be seen with the naked eye at first; but when once seen with the field-glasses, the mind's eye would always represent them to us, floating apparently skywards, in their vague and distant mirages. This discovery instantly created a burning desire in both of us, to reach them; but there were one or two questions to be asked before starting. We were now over forty miles from the depôt, the horses one night without water, and the distant ranges were sixty miles further away. There was no probability of water between, and we could not be sure if any existed in the new ranges. We might easily reach them, but if there was no water, how were we to get back? Close about, in nearly all directions except due north, were broken masses of hills; so I determined to search amongst them for water first. We therefore saddled up our horses, and departed for those nearest, which lay to the eastward; but in them we could discover no signs of the element we were in search of, nor any place, which would retain it for an hour after it fell from the sky. From here we went nearly north-west to a bare, red and inviting-looking hill, with cypress-pines on it, and here we were equally unsuccessful. From this we went nearly east again to some more pine-hills, but with like want of success, there being no place, where water could be expected to be found. From here we saw a kind of gorge or gap in the western portion of the bluff hill we had ascended this morning, and we now steered for it. It lay nearly south-west, but it being now night we had to camp, again without water, having travelled forty-six miles on all courses. The day was, comparatively speaking, cool, the thermometer standing at 90° in the shade.

Monday, 22nd December.—Thermometer last night 74°, this morning 62°. The night was again very cool, insomuch that I regretted bringing so small a quantity of bedding, for I was too cold to sleep. We were in our saddles again at five a.m., and steering for the gap we could not reach last night. Reaching it in six miles we found there was a small watercourse formed by a lot of little gullies. We traced it up into the hills, but every channel was dry. We then went over the ranges to the south, hoping to find a better locality on that side, but it was precisely similar to the north side, that is to say, there were little dry gullies running into an equally dry little creek. We next turned our attention to more pine-clad hills, lying nearly west, and arrived at them in eight miles, and renewed our search for water, but again without success. I did not now know, which way to turn; there was evidently no water whatever in any of these hills or ranges, and we were too far, to reach our depôt by night, so that the horses will be a third night without water—and very bad they are now for the want of it. Resolving to return (we however steered for another part of the range we had crossed in coming out) we continued travelling until we crossed our outgoing tracks, when we gave the horses an hour's rest. Starting again, and passing through the range more easterly than before, found another small creek, but as usual quite dry; so after getting clear of the stones of this range we steered straight for the valley near the depôt. We had to camp again without water, having travelled forty-two miles. Much time and distance was expended in our useless searchings for water. The day had been reasonably cool, the thermometer rising to 94° in the shade.

Tuesday, 23rd December.—Thermometer last night 80°, this morning 58°. We got back to the depôt in sixteen miles, and our horses at last got the water, they so much required. Gibson informed me, that during my absence the natives had been exceedingly troublesome, and had attacked the camp by throwing spears and stones down from the rocks above, so that he and Jimmy had been compelled to drive them away with firearms. Only about a dozen had shown themselves, though many others were seen trying to hide themselves behind the rocks, which are rather inconveniently situated, as they completely overlook the camp. It appeared also, that their friends had in my absence actually hunted all the horses away from their feeding-grounds, and neither Gibson nor Jimmy had seen them for three days, nor had they been up to water during that time. I did not however hear this, until I had let the three horses, which we had out with us, go. The remainder of the day we waited, hoping some of the others might come to water, but they did not. The thermometer indicating 98° in the shade at 4 p.m., we had our afternoon's earthquake, and Gibson informed me that twice during my absence they had occurred. The hostile action of the natives was the more annoying to me as it delayed me in carrying out my desire, to visit the distant ranges to the north. I had intended to have packed out some water fifty miles, and to leave Jimmy Andrews alone in the camp, while Mr. Tietkens and Gibson brought back a second supply for my horses on my return from the northern ranges, and it would actually necessitate, that only one should remain in the camp; but this news of the behaviour of the natives will compel me for a time at least to forego my trip. Gibson, I don't think, would stop by himself, though Jimmy expressed his willingness to do so, if I required it. I could not allow Mr. Tietkens to stop, as Gibson and Jimmy might go wrong with the water; so for the present I was compelled to relinquish my design. Christmas had been slightly anticipated by Gibson, who, when we returned, informed us that he had made and boiled a Christmas pudding, and that it was now ready for the table. We therefore had it for dinner, and did ample justice to it. Gibson and Jimmy had also shot several rock-wallabies, which abound here. They are capital eating, especially when fried like chops, when they have all the taste of mutton.

Wednesday, 24th December.—Thermometer last night 76°, this morning 70°. I sent Gibson and Jimmy out this morning, to look for the missing horses. Their usual beat is down the creek, and opposite Mr. Gosse's depôt. They at length returned, having found only the three horses Mr. Tietkens and I had out with us. The tracks of the others, and the tracks of a lot of natives driving them, were seen on Mr. Gosse's dray-track, so Mr. Tietkens and Gibson went on horseback after them, and it was late when they returned, having found them all. The natives had driven them several miles along the south side of this range, and they had got water in another small creek. While they were away some ducks visited our creek from the eastward; but, in consequence of the tea-tree being so thick, Jimmy and I were unable to get a shot at them. Thermometer to-day stood at 98°. Though the thermometer does not indicate very high, yet in this region there is a most oppressive weight in the atmosphere, and the thermometer seldom falls below its highest range until the sun is nearly down, so that all the afternoons are excessively relaxing. In consequence of the horses having once been hunted by the natives, and having found more water to the eastward, it is more than probable that they will never willingly water here any more, so from this time I keep one horse tied up at the camp, and will have all the others brought up every morning and taken back to their feeding ground.

Thursday, 25th December.—Christmas Day, 1873. Ah, how the time flies! What Ingoldsby says is, "Years glide away, and are lost to me—lost to me." Thermometer last night 86°, this morning 78°. While Jimmy Andrews was away (upon the horse, that had been tied up all night) after the others, we were startled out of our propriety by the howlings and yellings of a pack of fiends, in human form and aboriginal appearance, who had clambered up on the rocks just above the camp. I could only see ten or a dozen in front, but many more were dodging behind the rocks; the more prominent throng were led on by an ancient individual, who having fitted a spear was just in the act of throwing it down amongst us, when Gibson seized a Snider-rifle, and presented him with a conical Christmas-box, which smote the rocks with such force, and in such near proximity to his hinder parts, that in a great measure it checked his fiery ardour, and induced most of his more timorous followers to climb with most perturbed activity over the rocks; the ancient more slowly followed them, and then from behind his rocky shield he spoke spears and boomerangs to us, though he used none. He however poured out the vials of his wrath upon us—as he probably thought to some purpose; I am sure my interpretation of his remarks is correct, for he most undoubtedly stigmatised us as a vile and useless set of lazy crawling wretches, that came on hideous brutes of animals, being too lazy to walk like black men, and took upon ourselves the right to occupy any country or waters, we might chance to find; that we killed and ate any wallabies we happened to see, thereby depriving him and his friends of their natural and lawful game, and that our conduct had so incensed himself and his noble friends, who were now in the shade of the rocks near him, that he begged us to take warning, that it was the unanimous determination of himself and his friends to destroy such vermin as he considered us to be, and drive us from the face of the earth. It appeared to me, however, that his harangue required punctuation; so I showed him the rifle again, whereupon he immediately indulged in a full stop; they shortly after retired from those rocks, and recommenced their attack by throwing some spears through the tea-tree on the opposite side of the creek, when we had to discharge the firearms at them again. Just at this time Jimmy returned with all the horses, and we heard and saw no more of them. Jimmy brought up from Mr. Gosse's camp two pumpkins, which had been set there. One was half-eaten by wallabies, the other was whole; and they were most excellent eating, and quite a luxury to us. Gibson shot a fine wallaby, and we had fried wallaby-chops, pumpkins and melted butter for our Christmas dinner; and, drawing from the medical department a bottle of rum, we had a very excellent dinner indeed (for explorers); and having eaten our pudding two days ago, we had no occasion to envy any one their Christmas dinner (though we did). The day was very warm, the thermometer rising to 106° in the shade.

Friday, 26th December.—Thermometer last night 82°, this morning 72°. Nothing of importance occurred here to-day, but the temperature was much cooler than yesterday. Thermometer 90° in the shade.

Saturday, 27th December.—Thermometer last night 86°, this morning 72°. Having nothing else to do to-day I went out over the rocks wallaby shooting, but returned without success. Day warm, thermometer 102°. The rocks here get almost red hot from the fiery blaze of the sun, and the afternoons are most oppressive.

Sunday, 28th December.—Thermometer at night 90°, this morning 82°. The night was very warm, but a cool wind from the east rose after sunrise. The flies have become a terrible plague now, and especially to the horses, and more especially to the one, that happens to be tied up; one animal that was tied, when he found he could not break away, threw himself down so often and so violently, and hurt himself so much, that I was compelled to let him go. Thermometer 102°. A small fruit grows about here on a light green bush something like a grape, having one soft stone. They are black when ripe, and are very good eating raw; but we tried them as jam; and though all the others liked them, I could not eat them. This afternoon was very oppressive, and we had another earthquake shock, with a repetition of falling rocks and smashing trees.

Monday, 29th December.—Thermometer last night 92°, this morning 84°. The day was warm and a trifle cloudy. Thermometer 102° in the shade.

Tuesday, 30th December.—Thermometer last night 92°, this morning 78°. Night very hot in the early part, but a cool breeze from the eastward sprang up towards morning. Thermometer 102° in the shade. The water ceased running in the creek much earlier than usual to-day—namely between eleven and twelve; it usually ceases between two and three p.m.

Wednesday, 31st December.—Thermometer last night 88°, this morning 78°. The water ran again in the creek, soon after 10 o'clock p.m., and by morning everything was full again; though I fancy the current is getting a trifle weaker. Whether the moon has any effect upon this extraordinary water-flow, I cannot yet determine, but I fancy it has; I fancy as the moon waxes, the power of the stream wanes. Thermometer 102° in the shade.

Thursday, 1st January, 1874.—Thermometer last night 92°, this morning 78°. I am in great hopes of a fall of rain at the full moon. I have been most anxiously looking forward to it, as that seems the only time rain ever falls in this country.

Friday, 2nd January.—Thermometer last night 90°: this morning 80°; a very warm and unpleasant night. It became very cloudy at midday, and threatened rain from the east; at 2 o'clock a fine thunder-shower from that quarter came up, and, though it did not last very long, it quite replenished the water-supply in the creek, and set it running, after the stream had left off work for the day. Thermometer 94°, the sky cloudy; there was a sultry east wind blowing.

Saturday, 3rd January.—Thermometer last night 78°, this morning 78° also—a very cloudy morning. Yesterday's thunder shower has quite reanimated my hopes, and, in consequence of it, Mr. Tietkens and I again got our horses, and started off to revisit the hills lying north-easterly from here, hoping now to find sufficient water, to enable us to reach the more distant ranges, which we had seen from them. For the first ten miles from the camp we found the shower had extended, and the ground, in some places, soft, and we got sufficient water in a claypan to water our horses, had they required any; but, beyond that distance, no signs of rain having fallen were anywhere visible. The day was cool and cloudy, the thermometer going up to 96° in the shade. We went to the same place we had camped at on the night of the 20th December, being eighteen miles, and again without water.

Sunday, 4th January.—Thermometer last night 80°, this morning 82°—sultry night and cloudy morning. We took a very irregular course, going wherever we thought there was a possibility of a claypan existing, there being a few odd ones here and there in this part. At six miles we passed through a bit of scrub, and came upon a place, where there was a good-sized claypan, and which the natives had only just left, having evidently had long and numerous camps here; there was a clay-hole at one end of the pan, where water would lodge for some time after being filled; but the whole place was now perfectly dry. Continuing our searching march, we reached another kind of catching place in twelve miles. There seemed every likelihood of more rain falling, as thunder-storms were flying about thickly in all directions. We stayed here some little time, and dug out a clay-tank; while at it, a driving storm came up from the north-east, with heavy drops of rain. Unfortunately it lasted only half-a-minute; had it lasted only five minutes, it would have left enough water to serve my turn; as it was, it only wetted our clothes, and did no good to us. It was useless to go further, as there were no claypans at all further north, where we had been before, and I knew that no rains could have fallen, which would leave water in the hills, we had formerly searched. It was apparently evident, that if I wanted to reach the distant ranges to the north, I must pack out water, as Nature evidently did not intend to assist; though it seemed monstrous to be compelled to do so, while the sky was so densely overcast, with black thunderstorms coming up from all directions, and carrying right over our heads thousands of cubic acres of water, which it was reasonable to suppose must fall somewhere: so I determined again to wait a few days, and see the upshot of all these thunderstorms. To the east rains appeared to be falling all the morning; though, to the west, the horizon was as clear as possible. After finishing our tank here we returned to the native camp claypan, twelve miles back, hoping rain might have fallen there in our absence, as some of the storms had gone that way. It was late when we reached it, and it was as dry as ever; the horses had to go again without water. At dark another dense thunderstorm came up again from the north-east, and sprinkled us with rain for a few moments, and then travelled off south-west. The day had been very warm; while we dug the clay-tank, the thermometer indicated 104° in the shade. During the night we had plenty of thunder and lightning, but no rain.

Monday, 5th January.—Thermometer last night 78°, this morning 76°. The threatened rain has now entirely departed. The horses wandered away in the night, and it was late before I got them back to the camp. In the meantime Mr. T. had deepened the natural clayhole, that was here, making it able to hold double its original quantity. We packed up and started away again for the depôt, and found a little water about ten miles from it in a kind of little swamp. We turned the horses out, and dug another tank here, for the sky had again become cloudy. When we had finished our tank, we drained all the water in the little swamp into it; not having a pickaxe, we could only work slowly. I hoped this little reservoir would retain the water for a few days, as I intended from here to pack some water out. While digging, another thunderstorm came up, and split into halves, one going north, and the other south, apparently dropping rain on the country over which it passed—while we, as usual, got none. Saddling up again, we went home to await the awards of Providence. On nearing the depôt we found some rain had fallen here at all events, as some of the ground was a bit boggy, and there was water lying about. On reaching the camp, we were informed, that they had had two nice showers, and our water-supply was at all events replenished, as the stream now showed no sign of languishing, even in the heat of the day. Jimmy Andrews, with his usual brilliance, while we were away, had pulled a double-barrelled gun out from under a heap of bags and packs by the barrel; of course the hammer got caught, and snapped down on the cap, firing the contents—most fortunately missing Jimmy's body by only half-an-inch; had it been otherwise, we should have found him buried and Gibson alone. No natives had appeared while we were away. It was very hot at mid-day, the thermometer rising to 104° in the shade.

Tuesday, 6th January.—Thermometer last night 74°, this morning 74° also. In the middle of the night, my little dog Cocky rushed furiously out of the tent, and commenced barking at and chasing some animal round the camp, and eventually drove it right up into the tent. In the obscured moonlight, I supposed it was a native dog; but it was white, and looked exactly like a large fat lamb. When it sauntered away, I sent a revolver bullet after it—and hit it somewhere, as it departed at much better speed, squealing and howling until out of hearing. Just before daylight more thunder and lightning, with a few drops of rain; it seemed to be raining to the north. The sky was clouded all day; thermometer 96°. At night more storms were flying about, but no rain fell on us.

Wednesday, 7th January.—Thermometer last night 80°, this morning 68°. Mr. T. and I again got our horses, hoping at last—this being the third attempt, to reach the distant ranges to the north. It was late when we left the depôt, but we got to our old eighteen-mile camp, to the east of Mount Scott. While crossing the range before reaching the old camp we were met by a smart shower, which I felt sure would leave some water in our clay-tanks to the north; and we got plenty of water to-night to camp on. Thermometer has stood at 102° in the shade.

Thursday, 8th January.—Thermometer last night 74°, this morning 72°; a cloudy gusty night, with two light showers, but the morning broke fine and clear. Leaving the native claypan on our left, being sure it would now have some water in it, we pushed on to our outmost tank, nineteen miles from here, and were delighted to find that both it and the claypan near it were full, and I hoped the outside water would remain to water our horses on our return from the north. We remained here a couple of hours to rest our horses. It was a cool cloudy day, thermometer only 94°. We called this the Emu-Tank. After again starting we had another short shower, during which time the rain came down pretty heavily. We went on to the red hill with cypress-pines on, mentioned by me previously. We found a trickling flow of water of very small volume in a small gully, and there we encamped. I hope the water will continue to trickle until I return. From the summit of this hill, the centre of the distant ranges bore N. 10° W. They seemed very distant, with dense scrub intervening. We were now fifty miles from the depôt at Fort Mueller.

Friday, 9th January.—Thermometer last night 80°, this morning 76°; the night fine and clear. To-day for eighteen miles we found the scrub very dense, and at that distance turned our horses out for an hour. When we again started we entered on a mass of rolling open triodia-sandhills, thickly covered also with a most objectionable vegetable, which I always called sage-bush. In appearance it is (so to say) something like low tea-tree bushes, but it differs greatly from that shrub, inasmuch as it utterly abhors water. I have never mentioned it previously, though it grows almost everywhere in this region. I have collected specimens of it in my former and present expeditions, and though it is not spiny like the triodia, it is quite as annoying to both horse or man (if walking), as it grows too high for either to step over, and it is too strong to move easily aside; therefore a horse's track in this region is necessarily rather zigzaggy. At seventeen miles the open sandhills ceased, and more scrubs came on, but, having travelled thirty-five miles, we encamped upon the edge of it. Of course there was no water. The travelling was rather heavy. It was however a cool and cloudy day, the thermometer not rising above 96° in the shade. We passed through a few groves of the fine desert-casuarina, of which I am a great admirer, and which we have not seen for some time. A few native poplars and one or two curragong-trees were also met with to-day.

Saturday, 10th January.—Thermometer last night 84°, this morning 74°. The horses wandered back a long way in the night, and it was late before we started. At fifteen miles, now nearly approaching the range, we had been steering for, we debouched upon an eucalyptus-flat, which was covered with a most beautiful carpet of verdure. Not having met with gum-trees for some time, these looked exceedingly fine, and the bark dazzling white. We here found a crabhole with a little water in it, which our horses drank with great eagerness. The range was now only six or seven miles off, and stood up bold and abrupt, having steep and deep gorges here and there in its front. The range was timberless and whitish-looking, and I had no doubt of finding water at it. We stayed here a few minutes, to search for more water, as it would be an excellent place to return to to-night, if we could manage it, after visiting the new range. I was extremely annoyed here to find, that my field-glasses (an excellent pair) had been ripped off my saddle in the scrub, and I should now be disappointed in obtaining any distant view from the top of the new range when we ascend it; however, the glasses were gone—

> "They were lost to the view, like the sweet morning's dew;
> They had been, and were not, that's all that I knew—"

though of course I expected to recover them, when returning on our tracks. Starting again, we reached a gorge in the mountain's side at seven miles. Travelling through scrubs, over quartz, pebbly hills and occasional gum-flats—some with and some without creek-channels in them, all trending to the west, and, I supposed, joining and forming some important watercourse—we could discover a glittering little thread of water pouring down in a cascade from the top of the mountain into the gorge we were approaching, and upon reaching it found to our great delight, that we were upon the stony banks of a beautiful and pellucid little stream; so bright and clear was it, that at first I did not think there was any water there at all, until our horses splashed it up with their feet in crossing it. It was but a poor place for the horses to graze—not because there was no grass, but on account of the glen being so stony and confined. We turned them out here, and had some dinner and a most agreeable rest

"'Neath the gum-tree's shade reclining,
Where the dark-green foliage, twining,
Screened us from the fervid shining
Of the noon-tide sun."

This spot was distant 110 miles from my depôt at Fort Mueller. The thermometer stood at 96° in the shade; the day was, so to speak, cool and breezy. After our repast we walked up to the foot of the cascade along the margin of the transparent stream, which meandered amongst great boulders of rock. At the foot of the cascade we found, that the rocks rose almost perpendicularly from a charming little basin, into which the stream and spray from above descended with a most musical splashing sound, delightful indeed to an explorer's ear; the water fell from a height of 150 feet. The descent of the water was not quite unbroken, as the rocky face was not quite perpendicular. A most delightful shower of spray fell for many yards outside the basin, inviting us to a bath, which we most exquisitely enjoyed. The basin was not more than six feet deep; and I do not suppose this is a permanent cascade, but no doubt the water continues to descend for long periods after rain. I am quite delighted with this range, for it is new; no other traveller has ever been here. What a discovery! There were gorges to the right and left of this one, and no doubt each having plenty of water in them. At present I have no time to stay to explore them, but shall enter upon the task *con amore*, when I bring the whole party here. I hope to be able, from the features of this range, to pay a tribute to generous friends and supporters, by attaching their names to its points for ever. This little cascade however, I have appropriated to my only unmarried sister, and I have called it the Alice-Falls. It being impossible to ascend the mountain *via* the cascade, we had to flank it, and eventually reached its summit. The view from thence, though inspiriting, was still most painfully peculiar. Ranges upon ranges—some far and some near—bounded the horizon at all points from east round by north to west, though the western ranges did not appear to extend very far. A north line cut a great expanse of mountain country into halves; and, as far as the unassisted eye could determine, as much lay to the west as to the east, though nearer to me the eastern ones seemed the highest and thickest. There was a high, bold-looking mount or range to the north-west forty or fifty miles off, though others intervened much nearer. The particular portion of the range we were now on was 1000 feet above the surrounding level. I found the boiling point of water on this summit was at 206°, being the same as upon the summit of the Sentinel. The country intervening between this and other ranges seems open and good travelling country. The ranges beyond this have a brownish tinge, and are all entirely different from those at Fort Mueller. The rock formation of this hill, I am on now, is white and pinkish granite. All the ranges visible are entirely timberless; these are all more or less rounded and corrugated—some having conical summits and looking like enormous eggs standing up on end. I resolved to return at once, break up the depôt at Fort Mueller, and bring the whole party to this range. We therefore descended, caught our horses and departed, much gratified at the few discoveries, already made at this new geographical feature. We returned to the place, where we had got water in the forenoon, and encamped on the beautiful green grass-flat mentioned previously. The thermometer had only indicated 94° in the shade.

Sunday, 11th January.—Thermometer last night 80°, this morning 60°. A most delightful morning, like one of those delicious mornings one enjoys in the agricultural districts in the early spring. Our horses fed near us all night, and we got away early, travelling on to where we had outspanned in the middle of the day on our outward march, having found my field-glasses in the scrub. The day was delightfully cool; the thermometer never rising beyond 88° in the shade. We reached the red hill late, having travelled fifty-two miles. This enchanting spot was utterly and hopelessly dry, for the water had ceased to flow. I very foolishly, instead of pushing on to the Emu-Tank, went away to another place, about two miles off eastward, where we had been before, and I thought there might be water at it, but we were disappointed; so we had to camp there without water; the feed however was excellent.

Monday, 12th January.—Thermometer last night 80°, this morning 58°. The morning was so cold, that a full supply of blankets was agreeable. We got to the Emu-Tank pretty early. I was greatly surprised to find all the surface water was gone—our tank being the only place where any existed, and that was not full, a great deal having soaked away. Our three very thirsty horses greatly reduced its volume, and fearing it would all evaporate by the time we returned, we cut a quantity of bushes and sticks to cover it over with, though there was not sufficient water in it, when we left, for more than a dozen horses. Remounting our horses we went on to the native claypan, which we had avoided on our outward march. The claypan was almost dry; but the clayhole Mr. Tietkens had deepened, had a good quantity of water in it, but still not sufficient for all our horses when coming out this way; we therefore set to work and dug another tank, the soil being so hard, it all had to be chopped out with a tomahawk. We had to remain and camp here, as we had not finished it by night. The day was cool, thermometer 90° in the shade; but we found it quite hot enough working at the tank. I was greatly afraid the remaining shallow water in the claypan would be gone by the morning; but we went to sleep in hopes of its remaining at least for twelve hours longer.

Tuesday, 13th January.—Thermometer last night 80°, this morning 68°. We finished the tank early and drained all the water off the claypan into it—there being only just sufficient to fill it; then of course we had to cut boughs, bushes and sticks to cover it, and then we departed for the depôt. On reaching the ten-mile or kangaroo-tank, we found to onr disgust, that the water here was also nearly all gone, and our original tank was not large enough, so we were compelled to set to work and dig again at another tank, and when completed we drained all the remaining surface-water into it; then, of course, bushes, boughs and sticks had to be cut; and by the time we had completed everything, we were heartily sick of tank-making, our hands blistered, our arms stiff, and our whole bodies bathed in streams of perspiration, though the thermometer indicated only 98° in the shade. In consequence of having had to do so much tank-work, it was very late when we reached home. I was glad to find, that everything had gone smooth in my absence, and the natives had not been troublesome. We were even more pleased with the information, that Gibson had just shot a large wallaby, and it being cooked just as we arrived, we managed to devour every particle of it, as we had not tasted meat since we left on the 7th.

Wednesday, 14th January.—Thermometer last night 82°, this morning 74°. To-day we were getting all our packs and things ready for a start into the new regions, which we had so lately discovered, when, at eleven a.m., Mr. Tietkens gave the alarm, that the rocks overhead were lined with hostile natives, who, when they found themselves discovered, commenced a most direful yell. There were a great many more, than had ever formerly appeared. Of course they ordered us away and threatened us with their spears. They were also evidently in concert with others on the opposite side of the glen; but these reserves, although we could hear them, kept themselves out of sight. I waited for some little time to see, what they intended to do, when they became quite dauntless and descended from the rocks, intending evidently to wind up our little affairs at once. I soon became tired of inaction and their close proximity, and when they saw us move about under our bough-house they sent a quantity of spears into it, which it was miraculous we avoided. Mr. T. and I seized a gun and rifle and ran towards them, when they retreated towards the rocks, from whence they projected another flight of spears. We then fired at them, and when they found us becoming aggressive, they altered their tactics, and got down below us underneath the rocks. We however finally drove them from their stronghold, superinducing them to decamp, more or less the worse, and leave behind them a considerable amount of military stores in the shape of spears, wommeras, waddies, wallabies, owls, fly-flappers, red ochre and numerous other minor valuables, which we seized and brought in triumph to the camp. Their rout was complete; so we went on with weighing up our packs and getting all things ready for our departure. Three horses however could not be found, and we had to wait until the morrow. The day was excessively hot; the thermometer rose to 98° in the shade.

Thursday, 15th January.—Thermometer last night 88°, this morning 70°. The three missing horses were still away; so Mr. T., Gibson and I rode after them in different directions, but as only one (Gibson) found them, it was night, when I got back, and too late again to start; so I tied one of these wretches up, to get the others with in the morning. I was in a very uneasy state of mind about the water-supply in our tanks, as every hour's delay was of the greatest consequence. I had no great regrets at leaving this depôt, except that I had not been able to push more than 150 miles to the west from it. I now thought by going to the new northern ranges, that my further progress thence would be easier. We may perhaps have paid the passing tribute of a sigh at leaving our little gardens, for the seeds planted in most of them had grown remarkably well. The plants that throve best here were Indian gram, maize, peas, spinach, pumpkins, beans, cucumbers; melons also grew pretty well, with turnips and mustard. Only two wattles, out of many dozens sown, have come up, and no eucalypti have appeared, though many of several kinds were sown. Gibson had been most indefatigable in keeping the gardens in order, and I believe was really grieved to leave them, but the inexorable mandates of circumstance and duty forced us from our pleasant places, to wander into ampler realms and spaces, where no foot had left its traces. The day was warm; the thermometer had risen to 102° in the shade. When departing from here we left some lasting memorials of our visit. The water-supply was now better, than when we arrived—

"Our fount disappearing, from the rain-drop did borrow;
And to me came great cheering—I leave it to-morrow."

Friday, 16th January.—Thermometer last night 88°, this morning 70°. There was a cool breeze this morning from the east with clouds; there were a number of opossums here, which often damaged the garden produce in the night; there was also a most abominable croaking bird here at night, that used to annoy me exceedingly, and, though I often walked up the glen, I could never discover, what sort of bird it was; it might have been a raven. The horses were all mustered, and at length we were again about to depart—not certainly in the direction I should have wished, but still to something new. This was really as astonishing a place as it had ever been my fortune to visit; occasionally one would hear the metallic and silver-sounding concussion of some falling rock, smashing into the glen below, toppled from its eminence by some subterraneous tremor in the shape of an earthquake-shock. I cannot say, I should like to be caught in the act of wallaby-shooting amongst these rocks, while one of these trembling convulsions occurred, for the slightest vibration would precipitate the rocks into far different groups and shapes, than they before formed. Nor was this the only peculiar feeling, which I had with regard to this singular spot; for there was always a strange depression upon my spirits whilst here, caused partly by the knowledge, that it was still upon Mr. Gosse's track, and partly from the physical peculiarities of the region itself—

"On all there hung a shadow and a fear;
A sense of mystery the spirit daunted,
And said as plain as whisper in the ear,
This region's haunted."

At length we departed, leaving to the native owners of the soil this singular glen, where the water flowed only in the night, where the earthquake and the dry thunderstorm occurred every day, and turned our backs for the last time upon

"Their home by horror haunted,
Their desert-land enchanted,"

and plunged again into the northern wilderness. We reached our first tanks in eleven miles. The water had remained in them quite as well as I could expect; but as the horses had watered, when we left Fort Mueller, they required no more to-night. In consequence of the horses attempting to ramble back, we had to short-hobble them, but as soon as it was dusk they all went off at a gallop. Mr. Tietkens and I went after them, but the wretches would not allow us to get up to them, as they could hear us breaking sticks, while we traversed the scrub behind them, and as we could only follow them by the sound of their tramplings. We had to go five miles, before we got in advance, and, it being cloudy and dark, we scarcely knew where to drive them back to, amidst all thick scrubs; we fell over sticks, ripped our clothes and scratched ourselves almost to pieces, in trying to keep the brutes together. At length we saw the reflection of a fire, and it proved we were taking them right. It was nearly midnight, when we got back; we tied one up for security. Thermometer had stood at 102° in the shade to-day.

Saturday, 17th January.—Thermometer last night 76°, this morning 64°. A fine cool night, with cold wind from the east. In the morning of course all the horses were gone again, though we had even shorter hobbled them, when we brought them up at night, but having one to ride, the main mob was soon found; but it now appeared, that we had left three behind us in the darkness and scrubs last night, and although we searched in all directions they were not found, as their tracks had not yet separated from those of the main mob. We had now of course to water the horses at our tanks, and there was only just enough for them. The day was cool and breezy; there was fine green feed here; the thermometer did not rise higher than 90° to-day. I am getting very anxious about my middle and top tanks; the wretched horses causing so much delay is most annoying, the three missing ones could not be found by night.

Sunday, 18th January.—Thermometer last night 74°, this morning 60°. A cold and windy night; wind from east. Gibson and I went away again, and soon found the tracks of the missing horses, and very soon got them, and made a start from these now empty tanks at twelve o'clock. By night we got to the native claypan tanks. I was grieved to find, that the second tank we had dug here, though well covered with bushes, was quite dry; and the native waterhole contained only sufficient water for about half the horses. The thermometer rose only to 90° in the shade. Some of the horses soon drank all the water, and the rest had to go without, but I hoped to give them a drink at the Emu-Tank to-morrow, also to fill up our waterbags there.

Monday, 19th January.—Thermometer last night 82°, this morning 64°. Getting the horses early we got away by half-past six. On reaching the Emu-Tank, to my great disgust, we found it also perfectly dry, and were therefore unable to refill our bags, so we continued travelling all day as far as we could get. From here the country for twenty miles is dense scrub, so I had to ride ahead again with the bell on my stirrup; we got over forty miles to-day, and camped in the scrub without water.

Tuesday, 20th January.—Thermometer last night 74°, this morning 50°; a cold and delightful night. The horses rambled though short-hobbled, therefore we started late. Getting under weigh we pushed through more scrubs. At thirty-five miles we came to a spot, where Mr. Tietkens and I had seen water in a small Grevillea-watercourse; the surface-water had now disappeared; water however was obtainable by digging. The horses were excessively thirsty, and some of them upon obtaining a drink got the staggers. We encamped here, the grass being beautifully green; we crossed several flats with eucalyptus-timber all trending to the west, as I mentioned previously. The afternoon was warm, the thermometer standing at 102° in the shade.

Wednesday, 21st January.—Thermometer last night 74°, this morning 60°; a fine cool night. The horses having had two heavy days' stages, I did not move the camp to-cay; but Mr. Tietkens and I rode off to the Main-Range, to explore the gorges which we had formerly seen. Three of them lay to the eastward of the Alice-Falls; the most easterly, and which was near the end of that portion of the range, being about six miles from the cascade. We traversed very stony and scrubby country to the foot of the mountain, and reaching the mouth of the gorge, we tied up our horses and walked up. We very soon came up a fine deep long rock-reservoir, with the water running into and out of it. I could not touch the bottom with twenty feet of string, and I am sure the depth is over thirty feet. The rocky sides of this gorge rose almost perpendicularly above us, and the further we went up, the more water and reservoirs we saw, until our passage was completely stopped by the abruptness of the rocks and the depth of the basins at their feet. This is a permanently watered glen, and I have the pleasure to name it Glen-Cumming, after the Hon. John Cumming, M.L.C., Melbourne, a contributor to the fund for my expedition. The mountain in which this glen exists, and which forms the eastern extremity of this range, I named Mount Russell, after the Hon. Philip Russell, M.L.C., another Melbourne contributor to the fund for my expedition. Remounting our horses we proceeded westward along the foot of the range (towards the Alice-Falls); in a mile and a-half we came to another gorge, where there was another cascade, falling into a very fine deep round basin over twenty feet deep, and washed out of solid white granite. There were also numerous other rocky basins above and below the large one. I called this Glen-Gerald, after my young nephew, Gerald Buckland, of Geelong. Proceeding on, we came to another cascade and basin similar to the last, though the fall of water was from a lesser height. I called this Glen-Fielder, after my old friend Mr. Fielder King. From here we went to the Alice-Falls, and had a fine swim and shower-bath; there was a warm wind from the south-east blowing to-day. I wished in my travels to-day to find a road over or through these ranges, but there is no place near here to answer the purpose; I must go further along to the west, where at seven or eight miles there are apparently two separate hummocks; so I hope to be able to get to the northern side, between or around them. The day was warm enough, the thermometer standing at 100° in the shade; we returned to camp quite charmed with our day's ramble, though the country round the foot of the ranges is very stony and scrubby. I collected some stones from the hills; the vegetation about them is however in no way different to any, which intervenes between here and Mount Olga.

Thursday, 22nd January.—Thermometer last night 73°, this morning 66°; a cool and pleasant night. The horses having been let go yesterday without hobbles, had as usual walked off, and it was twelve o'clock before they were all mustered. Starting at length, we travelled about north-west, for the end of the most westerly of the two apparently separate hills. We passed through some scrub for a mile or two, and then came to some grassy eucalyptus-flats and small water-channels. At twelve miles we came fairly on to the banks of a splendid looking creek, with several fine sheets of water in it. Its bed was broad, and there were many channels, the intermediate spaces being thickly set with long coarse green rushes. The flow of water was to the northward, and the creek evidently went through a glen or pass. The timber on this creek grew thickly and vigorously; the water had a slightly brackish taste. All through the pass, which extended more than a mile, sheet after sheet of water was met with. We found one fine hole, with great quantities of ducks on it, but Gibson could not get his gun to go off, when he fired at them; so we were disappointed in getting any. We encamped at a waterhole near a very recent native camp, where the grass was most excellent, though we had not travelled a very long day's stage; but what of that?—I had discovered a fine permanently watered pass, and some excellent country round it to the south. The thermometer rose to 102° in the shade. Before entering this pass I found, that the line of ranges continued on unbroken to the west, and that this appeared the only pass through them. I named this "The Pass of the Abencerrages," that is to say—"The Children of the Saddle;" and called the creek and the sheets of water which exist here "Sladen-Water," after the Hon. Charles Sladen, C.M.G., of Mount Gellibrand, a contributor to the fund for my expedition. This evening, after having a comfortable bath and getting my blankets ready for bed, Jimmy Andrews came up to me and begged to know, what day of the month it was, and requested me to mention the fact in my journal, that Gibson was actually seen in the act of bathing! It was a circumstance well worth recording, for I am sure it is the first ablution he has indulged in since leaving Mount-Olga, and I am not sure that he bathed there—not but what he was particular enough in washing his hands and face, but it did not appear to strike him, that any other parts of his person required it; so with great pleasure I recorded the unusual occurrence.

Friday, 23rd January.—Thermometer last night 88°, this morning 74°; the night was warm, and the flies more than usually troublesome. This morning the latitude of this water I made 24° 29′ 10″, longitude 128°. Of course I shall follow this creek, to see where it goes. Making a start, we travelled along the banks of the creek, finding native gunyahs very numerous upon it, and for a few miles we saw several sheets of water; the bed then became too sandy. The main course of the creek was about north-west. In eight or nine miles we found that sandhills and casuarina-country existed, and finally swallowed up the unfortunate creek. The main line of ranges continued north-west or rather more westerly, and together with another range in front of us, to the north, formed a kind of crescent, and there appeared no pass through them. I now went to the eastern end of the range that lay to the north of us, and passed over a low ridge, and had a good view of the surrounding country beyond. Ranges appeared in almost all directions, the principal ones lying west and north-west. One conspicuous abrupt-faced mountain of some elevation bore north 17° east, which I named "Mount-Barlee," after the Hon. Frederick Palgrave Barlee, Colonial Secretary of the province, to which these ranges belong. There were others to the east-north-east, and the long sweep of the ranges we had now lately come from to the south. One hill on the range, I was now on, bore south 55° west, about four miles off. There appeared deep rocky ravines in its sides, and there seemed every probability of our obtaining some water there; so turning westward, along the northern slope of these hills, we reached a watercourse coming from the high hill before-mentioned; up it I found several very fine rock basins, ten or twelve feet deep. It was rather rough for the horses, but we encamped here nevertheless. I called the high hill above the camp Mount Buttfield, and the gorge a little to the east of it Buttfield's Gorge, after my very kind friend and most hospitable entertainer, J. P. Buttfield, Esq., S.M., Blinman. It did not appear, as if any rain had fallen here recently, because the water in all the holes was greenish and stagnant. The grass, what little there was, was old and very dry. The country consisted of open, sandy, level triodia-ground, dotted with a few clumps of casuarinas, giving a most charming appearance to the eye, but whose reality is startlingly different, keeping as it were the word of promise to the eye, but breaking it to the hope. Thermometer stood at 100° in the shade to-day. Our straight course was north 38° west, and we were only fifteen or sixteen miles from the pass; but having followed the creek, and travelled a roundabout road, we had travelled considerably over twenty miles.

Saturday, 24th January.—Thermometer last night 78°, this morning 63°. While the horses were being collected this morning I ascended Mount Buttfield, and found that ranges continued to the west for some considerable distance. A notch or fall appeared in the ranges, bearing 7° south of west, and I determined to make for it, as no doubt a creek issued from the hills there. The intermediate country consisted of some casuarina-sandhills and some level triodia-ground. Upon reaching the notch we found a creek with eucalypts; but it was quite evident, that none of the late partial showers had fallen here. There was little or no grass; the country at the immediate foot of the hills being stony, open and thorny. We searched about for water, but there was none whatever in the main-creek, and though there were several deep rocky basins, they were now all dry. Up in a small branch-gorge I was fortunate enough to find three small stony holes, with a very limited supply of water, altogether not sufficient to water the horses both now and in the morning, so they had to do without it to-night. They seemed willing enough, to eat such vegetation as grew here; but we kept them up in a kind of pound above the camp, where the hills on all sides fenced them in. The ants were excessively troublesome here. The thermometer stood at 102° in the shade, and it would be a difficult task to find any place, that was wide enough to screen any larger body than a thermometer. This glen or rather the vegetation, which had existed in it, had recently been burnt by the natives; this gave it a still more gloomy and dreary appearance. I called it by its proper name, that is to say, "Desolation-Glen."

Sunday, 25th January.—Thermometer last night 84°, this morning 72°. I could get no rest last night on account of the ants; the wretches almost ate me alive, and the horses tried so often to get past the camp in the night, that I was delighted at the reappearance of the morn. Mr. Tietkens also had to shift his quarters, and also drove the horses back, but no earthquake would ever rouse up the other two. Where the water was, was a most difficult place to get the horses to, and we could only take up three at a time. There was fortunately just enough water, and none to spare. One old fool of a horse must needs jump into an empty rock-basin, in which, being deep and funnel-shaped, he could not stand when he got there. He fell and knocked himself about terribly, before we could get him out. I never thought he would come out whole; and I was preparing to get him out in pieces, when he made one last super-equine exertion, and fell up and out at the same time. The delay in watering and extricating Terrible Billy from the basin made it twelve o'clock, before we could turn our backs upon this hideous glen, and I hoped we should not find many more like it. We travelled along the stony slope of the ranges nearly west. In less than two miles we crossed a small creek-channel, with a thick clump of eucalyptus-trees right under the range. The tops of another clump, half a mile off, were also visible. Mr. T. was away searching down Desolation-Creek. I directed Gibson to go on with the horses to the foot of a hill, which I pointed out to him, and remain there until I overtook them. I went up the little creek, and soon came to running water. Further up the whole glen was choked with a rank vegetation, beneath which the water ran along in a strong and permanent stream, which issued to the upper air from the bottom of the ranges. In trying to cross this watercourse my horse became entangled in the dense vegetation, whose roots, planted in rich and oozy soil, induced the tops of this peculiar plant to grow ten, twelve and fifteen feet high; it had a nasty gummy sticky feel when touched, and emitted a strong coarse odour of peppermint. I said in trying to cross this bed my horse became entangled in the vegetation, which was scarcely able to sustain his weight, and he commenced to plunge violently, and in consequence sank so much the deeper, that he precipitated me, head first, into the oozy black and boggy soil, and it was only the density of the vegetation, that kept him from being swallowed alive. I had in such a place great difficulty in getting my saddle, rifle, revolver and other things off the animal's back, and it was only with the most frantic exertions that I succeeded, though I gave up all hopes of ever recovering the horse, for he had ceased struggling, and was settling down bodily in the morass. I left him and ran off shouting and yelling after Gibson and Jimmy, but they never heard me, being now too far away; but most fortunately Mr. Tietkens riding past heard my calls and answered them; his astonishment was great indeed, when I showed him the horse so deep down in the morass; the vegetation held us up high and dry above the running stream, and at last by dint of kicking, beating, lifting and shouting at the creature, who, when he found we were trying to help him, commenced to help himself, he at length got out of this bottomless pit; he was white when he went in, but black when he came out. There were no rock-holes at the head of this glen; the water drains from underneath the mountains, and is permanent beyond a doubt. I called this "Luehmann's Springs," after Mr. Luehmann of the Botanical Gardens, Melbourne, who most kindly interested himself on behalf of my expedition. The water here runs but little over a mile. Re-saddling my dirty, muddy steed, we went on to the next gorge, where the clump of eucalypts was very thick and fine looking; the water here springs from the hills in the same way, as at the Luehmann's Springs, and we were pretty careful, how we approached with our horses this time; there was a fine stream of water flowing here. After this we found six other glens with running springs in about as many miles, they were named respectively (but afterwards) the next to Luehmann's Springs, "Groener's Springs," after Mr. Carl Groener, a faithful assistant of Baron Mueller's, and who took upon himself many onerous duties in furtherance of my expedition; the next one I named "Tyndall's Springs" after Mr. Tyndall of Black-Rock; the next I called "Fort McKellar" (where I subsequently had a depôt) after the Hon. Thomas McKellar, M.L.C., of Hamilton, a contributor to the fund for my expedition; the next I called the "Great Gorge;" the next the "Gorge of Tarns." All these places having permanent waters, though Fort McKellar is the most western water available for a depôt, and the most agreeable encampment. Many of these glens had fine rock-holes as well as running springs, and most of the channels were full of bulrushes, and that peculiar vegetation I mentioned before; it is of a dark green pulpy nature with a thick leaf, and bears a minute little flower of violet colour *(Stemodia viscosa)*. It seemed very singular, that all these waters should exist close to the place I called Desolation-Glen, where there was not a drop; and it seemed as though it was the only spot in the range that was destitute. After some time spent in exploring these charming glens and gorges, it was time to look about for the horses, for though Gibson had passed within sight of most of these waters, he had never stopped to see, if the horses would drink; however, we expected to overtake them in a mile or two, as the hill pointed out to him, was now close at hand. The country was so solid and stony, that one could not follow the tracks of the horses for any distance at a time, they could only be picked up here and there, but the country being open we thought to see them at any moment. and having found so many waters, and it being Sunday, I desired to camp early; but Gibson kept driving on, driving on, going in no particular direction, N., N.N.W., N.W., S.W., North again, so that having got such a start of us, it was just night, when we overtook him, still driving on up a dry creek going due south, slap into the range amongst rocks and stones, &c. I was greatly annoyed, for after having found eight splendid permanent waters, we had to camp without a drop either for ourselves or horses, the animals having been driven to death nearly, when we might have had a fine camp with green grass and splendid water. The thermometer had stood at 100° in the shade to-day.

Monday, 26th January.—Thermometer last night 88°, this morning 70°. A fierce south wind blowing all night. The ants were dreadful, and would not let me sleep a wink, though the others did not seem to feel them. The range still continued to run to the west, and there were other creeks visible; but I decided to return to the last water I had found, that is to say, at the Gorge of Tarns. I went after the horses very early this morning, and found they had wandered a very long way, although short-hobbled. I soon however found the cause; for one horse had been loose all night with his pack on, and had consequently led the others a fine jaunt. Having got them all, we returned to the last creek, which in consequence of its having so many splendid basins of deep water, I named as before said. On arriving we fixed our camp close up to the large basins, but the horses could water a mile below, where some teatree grew, and where the water reappeared upon the surface. This is the only creek on this range, where I have yet seen the teatree growing. There is a good bit of ground here for the horses to feed on, though not of too great an extent. There being a fine rocky tarn here, we were enabled to have as much swimming as we pleased, and even Gibson took a swim. Could the bottom of this pool be cleared of the loose blocks of stone, gravel and sand, it would doubtless be found of very great depth, but the floods of ages have nearly filled it with stones, loosened from the upper rocks, and it is only in the crevices between that one can discover the depth to be greater than seven feet; if cleared it might be seventy feet. The thermometer rose only to 98° in the shade to-day. The shade here is very scarce, except up around the large basin, where there are caves and overhanging rocky ledges, under which we sit.

Tuesday, 27th January.—Thermometer last night 88°, this morning 78°. A strong and warm wind blowing from the east this morning. Mr. Tietkens and I went up over the rocks, and found a great number of splendid tarns, some very deep, and water running from one to the other, and splashing over the rocks into the large basins below. We next climbed to the top of the range, but the view was almost similar to that from Mount Buttfield, only that now to the south we could see an horizon of scrub. To the north the view was bounded by a timbered horizon; but the natives burning the country in all directions, produced such a haze in the atmosphere, that no definite view could be obtained. To the west the view was impeded by other portions of this range; on this summit the boiling point of water was 206°. Not being able to glean any further information about the surrounding country, we descended to work in the shady caves, working and swimming alternately during the day, for we had plenty of the ever-recurring work to do, namely: the repairing of pack-bags and clothes.

Wednesday, 28th January.—Thermometer last night 90°, this morning 82°. A very close and hot wind all night. We remained here again to-day, as there was so much sewing to do. Our clothes require as much repairing as our pack-bags and saddles. No one could conceive the amount of tearing and patching, that is for ever going on. Could either a friend or a stranger see us in our present garb, our appearance would scarcely be thought even picturesque, for a more patched and ragged set of tatterdemalians it would be difficult to find on the face of the earth; not that we are actually destitute of clothes, but, saving our best for future emergencies, we keep continually patching our worst garments, hence our peculiar appearance. As our hats, shirts, &c., are here and there so quilted with bits of old cloth, canvas, calico, basil, green hide, and blankets, the original garment is scarcely anywhere visible. In the matter of boots, the traveller must be able to shoe himself as well as his horses in these wild regions of the west; and to give my readers some idea, what strength is required in boots for this country, I repaired mine with a double sole of thick leather, and sixty horseshoe nails in each boot, all beautifully clenched within, so that with an elegant corona of nail-heads round the heel, and toe-plates at the toes, they were perfect dread-noughts, and with such understandings I can tread upon a mountain with something like firmness. In the shade of the caves and over-hanging ledges the thermometer does not rise very high; to-day it only went up to 90°, but by eight o'clock at night, in the external air, it was no cooler.

Thursday, 29th January.—Thermometer last night 88°, this morning 76°. A hot and windy night, the wind blowing from the south. To-day we left this cool and shady glen—cool and shady however only amongst the caves—and continued our march still westward along the slopes of the range. The first eight miles of our course were 20° north of west, and in that distance we passed ten creeks issuing from gorges or glens in the range. All that I inspected had rocky basins with more or less water in them. From this point the range turned more westerly, still running on and throwing out more creeks; but no view could be got anywhere to the west, only the northern and eastern horizons being open to us as we travelled. On the northern side of the range, to the north, the country appeared to be open red sandhills, with casuarinas in the hollows between. The day felt very hot, though the thermometer only indicated 100° in the shade. At sixteen miles I found a large rocky tarn in a creek-gorge. There was little or no grass for the horses; but we had to camp there nevertheless. Indeed the whole country at the foot of this range is rather bare of that commodity, though at Sladen-Water the feed was excellent. But since leaving there, the horses have not done so well, and the country being so stony, most of them display signs of sorefootedness. I cannot expect that this range will continue further to the west than another day's journey; but I never can get any view in that direction. Many delays occurred in visiting numerous glens and creeks to-day, so that it was late enough, when we sat down among stones and triodia to devour our frugal supper below the rock-hole, where the water was. There was another large and semicircular basin still further up the creek which we called the Circus. This creek and the water-hole in it ever after went by that name in consequence of the singular basin I have just mentioned existing here.

Friday, 30th January.—Thermometer last night 78°, this morning 78° also; the night was cool and agreeable. Continuing our course this morning along the slopes, in a few miles I found that the range would now soon terminate; at nine miles (having crossed three creeks, which I did not inspect, as their gorges were some distance from our line), we came to a hill, disconnected from the main-range, and from it Mr. Tietkens and I obtained a view to the west, which consisted entirely of high red sandhills with casuarinas and low mallee. The view was closed by them at about ten miles. The long range we had followed so far was now at an end. It had fallen off slightly in altitude towards its western extremity, and a deep bed of sandhill-country, timbered with casuarina and covered with triodia, surrounded it on all sides. Nearer to us north-westerly, and stretching southward to west, was the dry irregular and broken expanse of an ancient lake-bed. Sweeping the horizon with the field-glasses, an outer and more distant horizon of precisely the same nature, though flatter from distance, was discovered; and at the point of about south-west by south, the top of a hill in a distant ridge was only just observable. We descended and rode on to the lake-bed, and found it very undefined. There were low ridges of limestone, with bushes and a few trees all over the expanse: the ground had patches of dry sodaish particles, and the soil was a loose dust-coloured earth. Samphire bushes also grew in patches, also patches of our enemy (triodia.) Great numbers of wallaby were seen amongst the limestone-rises, and they had completely honeycombed all we visited (hoping to find springs). But there was no water to be got anywhere. Returning to the horses we now turned southward towards an eucalyptus-creek, the most westerly that exists in this range. I rode on ahead, and found some water up at the head of it in some rock-holes; but it was so far easterly up the creek, that when we encamped we were not more than five or six miles from our last night's encampment across the hills, in a straight line; there was but a poor supply of water here, it all being contained in two small rock-holes, and it would not last longer than three days for my horses. The day was very hot, the thermometer rising to 104° in the shade. Some of the horses are now very footsore. I should shoe them, only the country surrounding this range being all sand, they will probably be off the stones again soon.

Saturday, 31st January.—Thermometer last night 90°, this morning 76°. I gave the horses a spell to-day, as I did not yet know, which way to travel next. Mr. Tietkens and I ascended the highest hill in this part of the range at the back of the camp, and from it we had a more extended view, than from the low hill we had ascended yesterday, though the characteristics were similar; the hills, whose tops I had then seen, were now found to form part of a low distant range, and not of a promising nature either. There was a very conspicuous mountain, which now bore from us to the north-east, between forty and fifty miles away straight, and a little to the west of it I fancied I saw refracted tops of other ranges floating in the mirage; of course I desired to go to the westward, and I thought from the mountain, just mentioned, I might discover others, which would lead me if not west at least north-west. Up to the present we had always called this mountain, in consequence of its original bearing when first seen, the North-west Mountain. I thought a change of country might be met with sooner in a north or north-westerly direction than west or south-westerly, as the sources of the Murchison-River must lie somewhere in the former direction. I tried the boiling point of water again here, and found the ebullition occurred at 2° higher than on the hill at the Alice-Falls, which indicated a fall of several hundred feet, and that this end of the range is much lower than the middle and eastern. My readers will form a better idea of this peculiar and distant mountain-range, when I tell them, that it is more than sixty miles long, averaging five or six miles through. It is of a bold and rounded form, there being nothing pointed or jagged in its appearance; its formation is granite, mostly white, though some portions are red; it possesses countless rocky glens and gorges, creeks and valleys, nearly all containing reservoirs of the purest water. It is a singular feature in a strange land; and God knows by what beady drops of toilsome sweat Mr. Tietkens and I wrested it from its former and ancient oblivion; its position in latitude is between the twenty-fourth and twenty-fifth parallels, and its longitude between 127° 30′ 0″ and 128° 30′ 0″, and I have the honour to name it the "Rawlinson Range," after that celebrated soldier, traveller, scholar and diplomatist, Lieutenant-General Sir Henry Rawlinson, President of the Royal Geographical Society of London. I found a singular moth-catching plant in this range, which exudes a gummy substance by which these insects become attached to the plant and die. A peculiar eucalypt, which grows most independently throughout the whole of this part of Her Majesty's dominions, was also here. I saw it first in the Glen of Palms and MacDonnell-Range of my former expedition. The bark is much whiter than that of the ordinary congeners, and the leaf looks different when seen on the tree, but when gathered and seen close no great distinction appeared to exist between it and the ordinary eucalyptus leaf. The general appearance of this range from a distance is white, flat, corrugated and treeless; it rises between eleven and twelve hundred feet in its highest points (about the centre, in the neighbourhood of Fort McKellar) above the level of the surrounding country, though its elevation above the sea is (206° boiling point of water) 3000 feet (approximately, as the exact height can only be determined by the means of numerous comparative observations.—F. v. M.)

Sunday, 1st February.—Thermometer last night 94°, this morning 78°. After a very hot night we made but a late start, in consequence of the trouble and delay in watering the horses, and a number of them were not satisfied, when we did leave; indeed these terribly rocky and stony places are most annoying sometimes both to men and horses, but where else in such a country can water be found? Our course was now for the North-west Mountain, which now lay nearly N.E. We had first to go some miles to get clear of the stones of this range. I had every reason to expect water at the mountain, to which we were now going, as its appearance was in every way most inviting, its corrugated front proclaiming the existence of ravines and gorges, while a more open valley ran between it and some lower hills showing themselves immediately to the west of it. The horses seemed so delighted, to get off the stones, that they travelled remarkably fast, and we managed to get over twenty-eight miles by night, though the country was exceedingly heavy travelling, being all high red sandhills, and until close at the end of our day's stage we could scarcely ever see the mountain at all. We encamped without water, but I expected to get some early to-morrow at the mountain. One or two of the horses lay down near the camp all night, being both thirsty and footsore, there being no grass for any of them. The day had been exceedingly warm, the thermometer standing at 108° in the shade.

Monday, 2nd February.—Thermometer last night 94°, this morning 80°. A great many of the horses, in consequence of being footsore, were lying down this morning when found, and they all looked excessively hollow and thirsty. I was forced to admit to myself, if no water was found at the mountain, that many of them would never return anywhere. We made an early start, and reached the stony foot of the mountain in twenty-three miles. There was a very inviting valley, up which we took the horses a mile; then leaving Gibson and Jimmy to await our return, Mr. Tietkens and I rode away in search of water. It seemed evident, that only a trifling shower, if any, had visited this range, for not a drop of water could be found, nor even any rock-holes where it might lodge. We parted company and searched separately, but when we met again we could only report to each other our non-success. By the time we returned to the horses it was past two o'clock. Our horses had been ridden rather fast for some time, over the most horrible and desolate stony places, where no water is, and they were now in a very exhausted state, especially Mr. Tietkens'. There were yet one or two ravines in the southern face of the mount, and while I ascended the mountain, Mr. T. and the others took the horses round that way, and searched again there. From the summit of this sterile mount I had expected at all events a favourable view of an horizon bounded, at least to the north, by near or distant ranges, but to my intense disappointment no such features presented themselves to my view. Two little hills only, bearing 20° and 14° west of north, were the sole objects higher than the general horizon, which was formed entirely of high red sandhills with casuarina between. To the east only was a peaked and jagged range, and which I called "Mount Robert," after my brother, Mr. Robert Giles, of Hamilton, Victoria. From the east-point, round by north and west and south of west, from whence we had come, the whole country was a mere bed of undulating red sand. What was to be hoped from such a region as this? Could water exist in it? It was scarcely possible. As for a creek or independent watercourse, I could not hope for such, because I had travelled so many hundreds of miles westward from my starting point at the Telegraph-Line, and no creek had ever been met except in the immediate vicinity of ranges; and not a drop of water, so to speak, had I obtained from them. These were my natural reflections on the top of this hill, which I was upon the point of naming "Mount Disappointment." As yet I had found no water at this mount, which had looked so inviting from a distance; and what possible likelihood was there of getting any in the midst of the dense bed of sandhills beyond? I did not test the boiling point of water here, for I had none to boil; but the elevation was about 1100 feet above the surrounding country. From a distance this mount has a very cheering and imposing appearance; and I would have gone almost any distance to it, with a full belief in its having water about it. But if indeed the inland-mountain has really voice and sound, all I could gather from the sighing of the light zephyrs, that fanned my heated brow, as I stood gazing hoplessly from its summit, was anything but a friendly greeting, it being rather a warning that called me away; and I fancied I could hear a voice repeating—

> "Let the rash wand'rer here beware:
> Heaven makes not trav'llers its peculiar care."

I now descended and joined the others at the foot of the hill. Mr. T. and Gibson informed me, they had searched all the ravines and water-channels, but entirely without success. The horses were huddled together in the shade of a thicket—three or four of them lying down with their packs on, and all looking pictures of misery. It was now past four o'clock, and there was no alternative but to retreat. The Gorge of Tarns was about the nearest water in the Rawlinson-Range, distant thirty miles, bearing south-south-west; but between us and it, was another low range, with a lower kind of saddle in the middle. I wished, if possible, to get over that before night; so we turned the horses' heads in that direction. Three or four of the horses were now in a very bad state; and I was afraid for one fine large harness-horse, called Diamond, who seemed suffering the most. Mr. Tietkens' riding horse, a small blue roan, a very game little animal, that had always carried him well, albeit not too well treated, was very bad. Two others also were very troublesome to drive along. The saddle of the low range in front of us turned out to be a much more difficult and stony pass, than I had anticipated. So dreadfully rough and scrubby was it, that I was afraid, night would descend upon us before we reached the southern side. Mr. Tietkens' Bluey here gave in, and fell heavily down a stony slope into a dense thicket of scrub. We had the greatest difficulty in getting him out; and it was only by sheer rolling him over the stones and down the slope, that we got him out at all. He got severely bruised in his descent; and when at the bottom he would not stand up. At length we got him on his legs and made him join the mob; and we just managed to get clear of the stones, when night came on, and we unpacked the exhausted animals, which had been travelling ever since daylight almost. We had of course no water for them, and none for ourselves. The day had been terribly hot, the thermometer had risen to 108° in the shade; ourselves and horses were choking for water.

Tuesday, 3rd February.—Thermometer last night 92°, this morning 84°. Several of the horses were lying dying about the camp this morning—Mr. Tietkens' Bluey, Diamond, a little cob and one or two more, while most of the fresher ones had wandered away. Though we were up and after them at three o'clock, it was ten a.m. before I could dispatch Mr. Tietkens and Jimmy with the main mob—poor little Bluey having fallen never to rise. Some of the horses were still away, and Gibson after them, while I waited at the camp for his return. He came at length and brought them, so we packed up and went after the others. Gibson's usual riding horse was very bad, and quite incapable of carrying him; my favourite little riding horse was by no means brisk, but still game. Mr. Tietkens was now riding an old horse, which I had purchased in Victoria, and have had for some time; he was called Widge; I had him out on my former expedition. He was a cool, calculating animal, that no ordinary work could kill; he was as lively as a cricket, when Mr. Tietkens rode away on him. Jimmy was mounted (and carried my little dog Cocky, now nearly dead from thirst, though we had given him our last drop) on a grey-hipped horse, which I had also brought from Victoria; he carried his rider well to the end. Gibson I had mounted on a young bay-mare, a creature as merry and gay as it is possible for any of her sex, even of the human kind, to be. Her proper name was the Fair Maid of Perth, but somehow, from her lively, troublesome and wanton vagaries, she was called the Sow-Cow. My own riding horse, a small sleek, cunning little bay, a fine hack with excellent paces, and which I also had out with me previously, I called West-Australian; but in consequence of his persistent attention to his *inwards*, he had received the soubriquet of "Guts." Poor Guts was now very bad and footsore. Gibson and I overtook Mr. Tietkens with the other horses, and we pushed on as fast as we could, the distance we had now to go not being more than ten or eleven miles. The sandhills were exceedingly high and severe, but all the horses got over the last one. We were now in full view of the range, with the Gorge of Tarns not more than five miles from us; but here Diamond and another horse (called Pratt) fell, never to rise. We took their packs off, and had to leave them on the ground. There was no shade, and the thermometer stood at 106° in such shade as could be got for it. We pushed on, intending to return immediately with water for the relief of these unfortunate creatures. The pack-horses now presented a demoralised and disorganised rout, travelling in a long, single file, it being impossible to keep the tail up with the leaders. I will give my readers some slight idea of them, if description is sufficiently palpable to do so. The real leader was an old black mare, blear-eyed from fly-wounds, for ever dropping tears of salt rheum; fat, large, strong, having carried her 180lbs. at starting, and desperately thirsty and determined. Knowing to an inch where the water was, on she went, reaching the stony slopes about two miles from the water. Next to her, a rather herring-gutted, lanky, bay horse, which, having been got from Mr. Bagot at the Peak, I consequently called Peveril; he was generally poor, but always able and willing for his work. Then came a big bay cob, an old flea-bitten grey (a fine harness-horse), and a nuggety black harness-horse called Darkie, always very fat. These last three I purchased from Mr. Gibson at Port Augusta; and capital horses they were, each carrying 200lbs. at starting.

Then Banks, the best saddle-horse I have, and which I had worked rather much in dry trips before reaching this range, and he was much out of sorts; then an iron-grey colt from Chowilla, called Diaway, having been very bad, when first purchased; then a sideways-going old crab of a horse called Terrible Billy. These three I purchased from Mr. Banks of the Burra. Terrible Billy was always falling into holes and up and down slopes, going at them sideways, without the slightest confidence in himself or apparent fear of consequences; the old wretch always did his work well though. Then came Blacky, a young and excellent colt, I got in Adelaide; then Formby, a North-West Bend horse, which I purchased from Mr. Formby in town, a horse that did me excellent harness work coming up to the Peak, always carrying top weight, 200lbs. at starting. Next was Hollow-Back, formerly belonging I believe to Dr. Blood, of Kapunda, a very game horse, though he only carried 180lbs. at starting. Then came Giant Despair—a perfect marvel of a horse—a chesnut, old, large-framed, gaunt and bony, with screwed and lately staked feet; life for him indeed seemed but an unceasing round of toil, but he was made of iron. No distance and no weight was too much for him; he sauntered along after the leaders, looking not a whit the worse, than when he left the last water, going neither slower nor faster than his wont. He was dreadfully destructive with his pack-bags; in timber he would never get out of the road of anything smaller than a big tree. Tommy and Badger, two of my former expedition horses, came next; Tommy, very footsore. Badger—a big, ambling cob, able to carry an enormous weight, but the greatest slug of a horse I ever met with, for he seems absolutely to require flogging as a tonic; he must be flogged out of camp and flogged into it again—mile after mile and from day to day, from water and to it. He was now as usual at the tail of a straggling mob, except Gibson's former riding horse, called Trew, as I purchased him of Mr. Trew of the Globe. He was an excellent little horse, but now so terribly footsore, that he could hardly drag himself along. Last horse of all was a little bay cob, the harness mate of the one killed and eaten at Elder's-Creek, on 22nd October. On reaching the stones, this poor little brute also fell, never again to rise; we could give him no relief, but pushed on. My little riding horse at length giving in, I let him go in the mob and walked to the water. I need scarcely say how thirsty we all were. On reaching the water and wasting no time, Mr. Tietkens and I returned as soon as possible to the three fallen horses with a supply of water, taking with us the old black mare, the Fair Maid of Perth, Widge and Formby; and we went as fast as the horses were able to go. On reaching the little cob, we found him stark and stiff, with his hide all shrivelled and wrinkled, mouth wide open, and lips drawn back to an extraordinary extent; pushing on, we arrived where Diamond and Pratt had fallen, they were quite dead also, they must have died immediately after they fell; they presented the same appearance as the little cob. Thus my visit to the North-west Mountain had cost the life of four horses, Bluey, Diamond, Pratt, and the cob; the distance was not great, which they had to travel, it being less than ninety miles, and they were only two nights without water; but the heat was intense and the country perfectly frightful, and to get over the distance as soon as possible, we travelled rather fast. The horses had not been well off for water or grass at starting, and they were mostly footsore; but in the best of cases and under the most favourable start from a water, the ephemeral thread of a horse's life may be snapt in a moment in the height of an Australian summer and in such a region as this, where that detestable vegetation, the triodia, and high and rolling sandhills exist for such enormous distances; the very sight of the country, in all its hideous terrors, is sufficient to daunt a man and kill a horse. I called the vile mountain, which had caused me this disaster, "Mount Destruction," for a visit to it had destroyed alike my horses and my hopes. I have the honour however to name the range, of which this desolate mount is the highest point, "Carnarvon-Range," after the Right Hon. the Earl of Carnarvon, Secretary of State for the Colonies, it being the most north-westerly point I reached upon this expedition. We returned again to the Gorge of Tarns, as Mr. Tietkens very tritely remarked, sadder but wiser men. Our position here is by no means the most enviable, for although there is plenty of permanent water at this range, it appears to be surrounded by such extensive deserts, that advance or retreat appear equally difficult, as now I had not a drop of water in tanks or otherwise, between this and Fort Mueller, and the question naturally arose, if I lost four horses in travelling ninety miles without water, how many should I lose in travelling 110 or 112 miles? Not a horse might ever reach it. I am now again seated under the splashing fountain, that falls from the rocks above, sheltered by the sunless caverns of this Gorge of Tarns, with a limpid and liquid basin of the purest element at my feet, sheltered from the heats and heated atmosphere, which almost melts the stones and sand of the country surrounding us, sitting, as I may well declare, in the shadow of a great rock in a weary land; but we cannot entirely shut out from the mind the perils already endured, and the perils yet likely to be endured. For the present our wants and those of our horses are supplied; but to the traveller in the wilderness, when he once turns his back upon a watering-place, the ever recurring question presents itself, when and where shall I obtain more? for the explorer is necessarily insatiable, no quantity will satisfy him, for he requires it always, and in every place. The thermometer to-day in the outer shade stood at 106°, in the caverns 98°.

Wednesday, 4th February.—Thermometer last night 96°, this morning 86°. The night was exceedingly warm and unpleasant. Of course we remained in camp here to-day, and we shall have to do so for some days, to recruit the horses. In the cool shades of the caverns and rocks the thermometer rose only to 98°, though it rose to 106° in the shade of a tree outside the glen. The bare rocks on either side of this gorge become so heated during the day, that it causes the nights passed in it to be most oppressive, as the rocks have not time to cool before the sun is upon them again; and at evening, when descending upon the caves, we find the thermometer often actually rises in the night air.

Thursday, 5th February.—Thermometer last night 101°, this morning 88°. Another hot and oppressive night, and though the flies are most troublesome, we are not plagued with either ants or mosquitoes. This evening Gibson and Jimmy shot three wallabies; this range swarms also with pigeons in every gorge or glen, and they come in clouds at night and morning for water; unfortunately nearly all our sporting ammunition is done, though I have a good supply of defensive. The thermometer in the caves to-day went up only to 88°; outside in the shade of a tree 104°.

Friday, 6th February.—Thermometer last night 98°, this morning 86°. A hot and gusty night, a hot wind from the south-east blowing all night. The thermometer in the caves rose to 96°. The external heat in the sun is most unbearable and oppressive towards afternoon. Clouds and the slight rumbling of thunder reverberated through the range to the east.

Saturday, 7th February.—Thermometer last night 100°, this morning 90°. Another hot and sleepless night. In the caverns the thermometer did not indicate above 90° all day.

Sunday, 8th February.—Thermometer last night 96°, this morning 92°. We are still resting the horses here. The daytime in this peculiar gorge is less oppressive than the night, as the sun does not appear over the eastern hills until nearly nine o'clock in the morning, and it passes behind the screen of the western ones at 4.15 p.m. There was a good deal of lightning seen flashing above the cliffs last night, but no clouds or thunder. Thermometer in the caves to-day 90°.

Monday, 9th February.—Thermometer last night 98°, this morning 92°. A close, hot and sultry night. Shod a couple of horses to-day which were very tender-footed. To-morrow I intend to return to Sladen-Water, as that is the only place about this range, where grass, water and soft ground for the horses can be got together. My little dog, in jumping about my bed this morning, broke one of my thermometers. It was cloudy to-day with some thunder, and twenty drops of rain fell in the afternoon.

Tuesday, 10th February.—Thermometer last night 96°, this morning 86°. A rather cooler night than we have had for some days. The day however was hot enough, thermometer in the outer shade 104°.

Wednesday, 11th February.—Thermometer last night 98°, this morning 88°. Getting the horses up, we started away from this friendly gorge, and followed down the range to the eastward. One horse (Tommy) was still very bad, so we did not go very far, only ten or twelve miles, and got some water by digging in a creek east of Desolation-Glen about five miles. Thermometer 104° in the shade.

Thursday, 12th February.—Thermometer last night 98°, this morning 86°. Started again this morning, hoping Tommy would travel to the Pass of Sladen-Water. At ten miles however Tommy gave in, not from thirst, but from exhaustion, I suppose. We left him in the sandhills under a casuarina-tree. Pushing on we arrived at the Pass early. The sheets of water in it were now not of quite such large dimensions, as when we first saw them, and it had become more brackish. Most of the country travelled over to-day consisted of triodia-sandhills. We had now come through a kind of open pass between this range (the Rawlinson) and the Mount Buttfield Range. Formerly we had gone more easterly *via* Mount Buttfield itself, and thence westerly. There was a fine flock of ducks at the waterhole, we now encamped at, and Gibson and Jimmy went blazing away as usual without even getting one, wasting powder and shot, which has now become quite a scarcity with us, and making the ducks wild into the bargain.

Friday, 13th February.—Thermometer last night 92°, this morning 82°; a reasonably cool night. I sent Mr. Tietkens and Gibson back to-day to try and recover Tommy, they taking a supply of water for him. I rode up through the Pass and searched for a better-tasting water than that of our present encampment. One fine sheet was quite good, that is to say it was very sweet, and made excellent tea without sugar. The ducks had been so frightened by Gibson and Jimmy's useless firing, that they split into several mobs, and only one mob of eight remained at our camp. I had some difficulty this morning with these, as I went and got a horse and drove them down to our waterhole, to get a pop at them; but of course the moment they arrived, I being behind with the horse, Jimmy must needs go blazing at them again, although he knew that he couldn't possibly hit one; and just when I arrived I only heard the report and saw all the ducks come flying overhead again up the pass. However a few hours afterwards I got a shot at them, and by good fortune bagged six, and one got away in the rushes. They were the fattest birds I ever saw. Thermometer to-day 102°. Mr. Tietkens and Gibson returned at night, having got Tommy standing precisely where he was left, and they brought him on to the furthest water down the creek, as they had a great deal of trouble in getting him so far. We had a fine supper off the ducks, their taste and flavour being super-excellent. During the afternoon we had some thunder and lightning and a few drops of rain for about half-a-minute, when the sky became perfectly clear again.

Saturday, 14th February.—Thermometer last night 90°, this morning 80°. More thunder and lightning were both seen and heard, principally to the east and north-east. The ants were most terribly troublesome at this waterhole, although we were encamped upon the damp sandy bed of a portion of it. We shifted the camp this morning up to the sweet waterhole, and selected as open a piece of ground as possible, as I intended the camp to remain here for a week or two at least, as I wished to visit the low ridges I had seen south-westerly from the west end of this range, before we visited Mount Destruction. We had more thunder and lightning to-day, the sky being greatly overcast; a few drops of rain only fell. The day was very warm, the thermometer rising to 106° in the shade. At night vivid and continuous lightnings appeared in the north. There are countless numbers of the little cockatoo-parrots here; they are rather shy, but when a gun is let off at them, even by Gibson or Jimmy, a dozen are sure to fall; but it takes some time before another shot can be got at them. The pigeons swarm every night to water; the parrots, I fancy, are migrating.

Sunday, 15th February.—Thermometer last night 90°, this morning 82. A warm, windy night; wind from the south. It was also windy and hot all day, though the thermometer rose only to 100° in the shade.

Monday, 16th February.—Thermometer last night 80°, this morning 70°. A cool but very windy night, with a dull and cloudy morning; wind still from south. To-day we shod the old black mare, as I intend taking her with me to the south-west. The day was comparatively speaking cool, but windy, the thermometer rising only to 90° in the shade.

Tuesday, 17th February.—Thermometer last night 80°, this morning 68°. Nearly all the horses are away, the creatures having found another drinking place in a small branch creek, two and a-half miles off to the south. Though there is most excellent grazing country here for the horses, it is excessively scrubby, which makes it very difficult to find them. The day was cool, the thermometer rising only to 86° in the shade. We shod Diaway, the Chowilla colt, to-day, which I am also going to take with me this time to the south-west.

Wednesday, 18th February.—Thermometer last night 72°, this morning 72° also. This was like to have proved a most eventful day in my life, as it was near the termination of it; for I was riding Diaway, the colt we had shod yesterday; he is seldom ridden (though a fine hack) as he is such a good weight-carrier, and he is a rather skittish brute as well, and if anything goes wrong with his pack, he'll put it right almost instantaneously. I was driving all the horses towards the camp, when one broke from the mob and went over the creek; there was a bank of stones about three feet high on the opposite side, which was almost hidden by the growth of rushes. Diaway went bounding over the great bunches of rushes and reached the bank without seeing it, until too late; he made a bound at it, but fell on top of it, and on top of me also, leaving me on the broad of my back. On my feet were those wonderful boots, before described, with the sixty horseshoe nails in each, and it was no wonder, that my foot got caught in the stirrup on the off-side of the horse. Reader, were you ever dragged by a horse? It is one of the most horrible positions to contemplate, that the mind can well imagine. I have been dragged four times before now, and have only escaped by miracles. In this case the horse, finding me attached to him, commenced to lash out his newly-shod heels at me, bounding away at the same time into the thick scrub, which existed close here. Mr. Tietkens and the others, seeing the accident, came running up behind, as we departed. I was fortunately not dragged far, but was literally kicked free by the impatient and now frightened animal; the continual kickings I received, some on my legs and body, but mostly upon that portion of my frame, which it is considered equally indecorous to present either to a friend or an enemy, at length bent one or two of the nail-heads, which held me and tore the upper leather off my boot, which fortunately was old, and ripped it clean off, leaving me at length free; as I lay on my back I saw that animal depart without me into the scrub with feelings of the most profound delight, though my transports were considerably lessened by the agonising sensation I experienced in my rear. Mr. Tietkens helped me up, and I hobbled up to the camp in a most disorganised state, though thanking Providence for so fortunate an escape; for had the horse entered the scrub, not two yards from where I became released, I could not have existed more than a minute. The day was most agreeably cool, the thermometer rising only to 88° in the shade.

Thursday, 19th February.—Thermometer last night 80°, this morning 70°. This morning Mr. Tietkens was getting everything ready for our departure for the S.W. ridges (though I had great doubts of my ability to ride), when we became aware of the presence of a whole host of natives immediately below the camp; the little dog had been strongly perturbed for some time previously. We knew the natives were in our vicinity, as they had been burning the country all round for the last two days; there being so much long grass and rushes in the creek-bed, the wretches could approach very close to us, before either we or the little dog could see or smell them. As soon as these individuals found themselves detected, they (as usual) set up the most horrible yells and shrieks, and came running up towards us, waving their shield and spears; five or six in the front rank, more energetic than the others, had fitted their spears for throwing as they ran, and before a gun could be seized, five spears came rattling into the camp, and we had to jump behind a large bush that was there to escape them. Mr. Tietkens soon seized a rifle, and sent a Snider-bullet at them, it flying just above their heads—the report of the weapon and the whirring sound of the swiftly passing ball instantly made them pause, and they commenced an harangue and ordered us away out of their territories to the south; but finding us immovable and speechless, their courage returned, and again they advanced, uttering their war-cry with renewed energy. Again the spears would have been amongst us, only that I determined not to permit either my own or any of my party's lives to be lost for the sake of not discharging our firearms, consequently we drove them off, and caused a complete rout amongst them. There was a great number of them, though we had to do with only twenty or thirty, the rest were hid behind bushes and rushes; there were not many oldish men amongst them, only one with greyish hair. I am reminded here to mention, that I have seen no such things in any of my travels in these western wilds as graves of any kind. It is not, that they are consumed by the continual fires the natives make use of here in hunting, as that would likewise be the fate of their old and deserted gunyahs, which we meet with frequently, and which are not all, nor half consumed by fire. My opinion of these people is, that they eat their aged ones, and most probably eat those who die from natural causes also. The day was breezy, cool and agreeable, the thermometer only 92° in the shade. In consequence of the natives having attacked us, Mr. Tietkens and I did not depart on our S.W. excursion. I felt pleased enough to delay my departure, for I was in great pain all over from the accident of yesterday. In consequence of this hostile action of the natives, I decided to desire Mr. Tietkens, to remain in the camp with Gibson, while I took Jimmy Andrews out with me. I cannot say I am in great hopes of making much of a discovery upon this next trip I am contemplating; for had there been any ranges of any elevation to the westward beyond the ridges I am now going to, I should in all probability have seen them from the end of this range, but all I saw were some low ridges to the S.W. I felt it incumbent on me to visit these ridges, as from them I might obtain a view of some other features beyond.

Friday, 20th February.—Thermometer last night 80°, this morning 72°. Taking Jimmy Andrews with me to-day we left the camp at Sladen-Water reasonably early. The hills I was now about to visit were, as said before, not very inviting, as from the western extremity of this range I had turned towards Mount Destruction in preference. I had however been there, and it was necessary to visit these others also. This time, after debouching through the pass on to the south face of the range, I travelled along its foot westward, crossing several small creeks, which might or might not have had good gorges in them; but as I now kept some little distance from the main-hills, in consequence of the roughness of the ground, I did not inspect any of them. At twelve miles we came to a small and very green water-channel, and found water running down it, as far as a rocky hole close to where we crossed it, and where it then sank below the surface. We turned our horses out here for an hour, though we had come so short a distance; I found riding very severe exercise after my late accident, and was only too glad to rest. I named this agreeable and pretty little creek "Glen-Helen," after my niece, Miss Buckland, of Geelong. Starting again we kept as near the ranges as the nature of the ground would permit, but it was exceedingly rough travelling, far more so than on the northern side of the range; at three miles we crossed another running water in a finer and larger channel than at Glen-Helen, and there was much more water. I called this "Edith Hull's Springs," after my sister's step-daughter, Miss Hull, of St. Kilda. We did not stop here, but continued our course. At ten miles further, after crossing several other creeks, whose heads all looked more or less inviting, we turned up one creek, and got some water in a very rough and stony gorge, off the main channel, which was dry. There was very poor feed indeed for the horses, and though they were shod, the place was so terribly rocky, that they preferred to stand still rather than attempt to roam about; the vegetation was nearly all triodia. The day had been reasonably cool, the thermometer standing at 92° in the shade at Glen-Helen, though I think it would have gone higher after we left that place; the horizon to the south, the only one we could see, was bounded entirely by red sandhills and casuarina-timber.

Saturday, 21st February.—Thermometer last night 82°, this morning 72°; the horses ate nothing all night, and had stood almost where they had been hobbled. The moment horses, however fat and fresh they may be, are taken away from their mob of companions, to face the fearful country, that they know by instinct is before them in these regions, during the summer heats, they begin to fret, and fall away visibly; they will scarcely eat and in consequence get all the weaker, and then they require twice as much water as they otherwise would, if their insides were partly filled with grass; when I released them from their hobbles, they immediately commenced to feed; but of course they had postponed their meal rather too late, as it was now time to travel. They were very thirsty again this morning, when we took them to the rock-basin, where we were able to indulge in a bath before starting. Leaving this camp we travelled onwards still farther west, the foot-hills, amongst which we were obliged to travel, being much rougher than even yesterday's road, so much so that I kept away as much as possible, until we had gone twenty miles; it was then time to look out for a watering-place; and going up a most dreadful gorge, choked up with huge boulders of red and white granite, I at length found a fine rock-tarn, indeed I might well call it a marble bath, for the rock was almost pure white and perfectly bare all round about; the water was considerably over our heads, and felt as cold as ice; it was a most dreadful place to get horses up to, and two of them fell two or three times on the slippery rock; one old horse (the one I had nearly lost in the morass at Luehmann's Springs, on the other side of the range) hurt himself a good deal. It was late in the day, when the horses had been watered, so I determined to camp here, though the grass was very little better than at our previous night's encampment, nor was the glen any less stony or rough. The day was 12° hotter than yesterday, the thermometer rising to 104° in the shade. The ants in this glen were perfectly frightful, they would not allow me to rest a moment anywhere. There was but one eucalyptus-tree, and in the shade of that they swarmed in countless thousands; the sun poured his beams full down upon us, and it was not until he departed over the cliffs, at 5.30 p.m., that we had a moment's peace; indeed I passed the time mostly in the marble-bath and then took a walk up to the top of the range. The south-western hills I desired to visit, bore nearly in that direction. I determined not to go any further along the range, but start for them from here. As long as the sun was pouring his rays down upon their unsheltered hides, the horses would not attempt to eat; but when he departed, they fed a little on the coarse vegetation, which existed here. This glen, like all the others in this range, swarmed with pigeons; we got some for breakfast, enough fortunately at one shot. During the hot months I believe whites could live entirely on pigeons in this range. At the camp at the Pass they come to water in clouds, and we have been living entirely on them, for now we have no other meat. But unfortunately our ammunition is nearly all but gone; as long as it lasts we shall have birds; when it is gone, we must eat horse-flesh (and should have done so for some time only for these birds). I have an old horse now fattening for the knife, and I am happy to say, whenever I inspect him, he appears to look better each time. The animal I mean is the old sideways-going Terrible Billy: poor old creature! to work so many years as he has done for man, and then to be eaten at last, it seems a hard fate; but who or what can escape that inexorable shadow? It may be the fate of some of us to be eaten, for I fully believe the natives of these regions look upon living organisms as grist for their insatiable mills. As night came on I was compelled to lie down at last, but was so bitten and annoyed by the ants, that I had to keep moving from place to place the whole night; but the (in) sensible Jimmy lay sleeping and snoring (though swarmed over and almost carried away by the ants) as peacefully and calmly, as though he had gone to rest under the canopy of costly state, and lulled with the sounds of sweetest melody. I could not help moralising (as I often stood near him, wondering at his peace and placidity) upon the difference of our mental and physical conditions: here was one human being, young and strong certainly, sleeping away the dreary hours of night, regaining that necessary vigour for the toils of the coming day, totally oblivious to swarms of creeping insects, that not only crawled all over him, but kept constantly biting into his flesh; while another, who perhaps prided himself too much upon the mental gifts bestowed by God upon him, was compelled by the same insects, to wander through the whole night from rock to rock, from place to place, unable to remain for more than a moment or two anywhere, and to whom sleep under such circumstances was an utter impossibility: not that the loss of sleep troubles me at all; for if ever there was a mortal, who could claim the title of "the sleepless one," it would be me; that is to say: "when curtained by night and the stars;" and though I do not sleep much, I still require a certain amount of horizontal repose, which in this fearful glen I could not obtain. It was therefore with extreme pleasure I beheld—

> "To the eastward, where cluster by cluster
> Dim stars and dull planets that muster,
> Waxing wan in a world of white lustre
> That spread far and high."

Sunday, 22nd February.—Thermometer last night 92°, this morning 78°. Very few human beings could have been more pleased than I at the appearance of this day, though I was yet doomed to several more hours' misery in this dreadful gorge. The pigeons, shot last night, were covered within and without by ants, though they had been put in a bag. The horses looked wretched; after watering them, they seemed inclined to feed; so I determined to give them a day's rest, before I ventured with them into the frightful sandhills, which intervened between us and the distant ridges. Truly the time which I spent in this gorge was hours of torture, the sun pouring down his fiery beams again upon us, the rocks and stones all becoming heated, and the ants more numerous than ever, so that I almost cried aloud for the mountains to fall upon me, and the rocks to cover me. I passed several hours in the marble-bath, as being the only place, where the ants could not come (but there were numerous little water-beetles, that were almost as bad) until I became quite benumbed with cold. Then the sunshine and heat in the gorge would seem quite delightful for a few minutes; the thermometer stood at 106° in the shade of the only tree at 3 p.m., when the horses came up themselves to water. I was so horrified with this place, that I could no longer remain, though Jimmy sat and probably slept under the one tree's scanty shade, and seemed to pass the time away as comfortably as though he were in a fine house. In going up to the water two of the horses again fell and hurt themselves. But the old blear-eyed mare she never slipt or fell, She would take a week in the passage rather than fall. At 4 p.m. we mounted and rode down the glen until we got clear of the rough hills, when we turned upon our proper course for the ridges, we were about to visit, but which we could not see. In two or three miles we entered the sand-hill region once more, though for a few miles it was somewhat flat, but rose into hills at length. The triodia was as thick as it could grow; but the country was not to say scrubby, there being only low bushes and shrubs on the sandhills, and casuarina-trees of beautiful outline and appearance in the hollows. When the horses got clear of the stones, they commenced eating everything they could snatch and bite at, as we travelled along. Finding a patch of dry grass amongst some clumps of casuarinas at fifteen miles, we encamped, of course without water. The horses fed pretty well for a little time, until the old mare began to think it time to be off, and she soon would have led the others back to the range, for she dreaded this country, and knew what agony was in store for her by experience and instinct, and therefore omitted no opportunity of leaving her masters in the lurch, if allowed. Jimmy got them back, and we very short-hobbled them. The ants were annoying here, but nothing to be compared to those at the gorge we had left, and I only had to remove my bed three times during the whole night, so that I had a most delightful night's rest, though of course I did not sleep.

Monday, 23rd February.—Thermometer last night 86°, this morning 60°. The night and morning were remarkably cool. The horses had gone as far as they well could during the night with the short-hobbles they had; but they had got sulky, and refused to eat anything, and this morning they looked as hollow as drums, and seemed totally unfit to traverse the ground that was before them. However, it had to be done, or at least attempted, and we got away from camp at an early hour. We were already in the thick of the sandhills, and here they rose almost into mountains at intervals. It was most fatiguing to the horses, the day being quite too hot. At twenty-two miles I gave them a rest in the umbrageous shade of a very large curragong-tree, when the thermometer rose to 104° in its shade. There was no grass, and had there been the animals would not have attempted to eat; so they merely rested. We had only within the last mile or two sighted the hills, I had been steering for, not having seen a vestige of them for thirty-seven miles. The first hills were now pretty close to us; they seemed very low and quite unlikely to supply any water. Resuming our march, we reached the first hill in eight miles, and upon my ascending its summit I could see at a glance, that any prospect here of finding water was utterly hopeless, as these low ridges, which ran north and south, were merely a few oblique-lying layers of upheaved granite, not much higher than the sandhills which surrounded them. There were a few more disconnected hills belonging to this group lying north-west, west and south-west, all within a circle of eight or ten miles. There was literally no place, where water might be got even during rains. The whole country round was simply a dense bed of rolling sandhills. Not a rise could be discovered in any direction, except of course that from whence we had come. I descended to Jimmy and the horses, and we went on west to the furthest hills connected with these miserable ridges. We got to the foot of the most westward of them in six miles just about sundown, and long will that peculiar sunset rest in my recollection. It seemed so strange and so singular, that only at this particular sunset out of the millions, which have elapsed since this terrestrial ball first floated in ether, I or indeed any white man should stand upon this wretched hill so remote from the busy haunts of my fellow-men. My speculations upon its summit—if indeed so insignificant a mound can be said to have a summit, were as wild and as incongruous as the regions, which lay stretched before me. In the first place, I could only conclude, that no water could exist in this region, at least as far as the sand-beds extend. I was now (though of course I had come a great deal to the southward also) nearly thirty miles to the west of the most western portion of the Rawlinson-Range. From that range no object had been visible above the sandhills in any westerly direction, except these ridges I am now upon, and from these if any others or any ranges of higher proportions existed anywhere within one hundred miles, I must have sighted them; and the inference to be drawn in such a case was, that in all probability this kind of country would remain unaltered for an enormous distance, possibly to the very banks of the Murchison River itself; and the question naturally occurred to me, could it be penetrated by man with only horses at his command, particularly at such a heated time of year? Where was Colonel Warburton, and his party? Had he also met with such a desert? Could he, even with his camels, penetrate it? I wondered where he could be. Had he been successful, for I never doubted his high courage? Had he returned? Or could he ever return? Mr. Gosse and his party I had every reason to suppose would return in safety; but let me assure my reader, that his undertaking was indeed great, and his retreat most decidedly was wise. I only coveted from my brother explorers one thing, and that was their camels. What are horses in such a region, in such a heated temperature as this? The animals are not actually physically capable of enduring the terrors of this country. I was now scarcely one hundred miles from my main camp, and the horses had had plenty of water up to nearly half-way; now they looked utterly unable to return. What a strange maze of imagination the mind can wander in, in recalling the names of those yet-separated features, the only ones at present known to supply water in this region, that is to say: the Murchison-River and (my newly-discovered) Rawlinson-Range. The Rawlinson and the Murchison, named after two Presidents of the Royal Geographical Society of London, the late and the present, both most eminent and renowned—the living and the dead—physically and metaphysically also, are not the features as the men separated, by the great gulf of the unknown, by a stretch of that undiscovered country, from whose bourne no traveller returns? The sun went down, and I returned to my youthful companion, who was below with the horses. We were fifty-one miles from the water we had left. The horses looked perfect pictures of misery. Old Buggs's (the grey horse) legs had swelled greatly from the contusions he had received in falling, while going to and from the water. The old black mare, which I had been riding, though but a sorry hack, looked very bad also. Even the youthful and light-heeled Diaway hung his head, and you could almost span him round the flanks; their miserable appearance was caused as much from want of food as want of water, as they had scarcely eaten a toothful since leaving Sladen-Water; indeed all they had seen to eat was not inviting. We mounted and slowly left these desolate ridges behind us, and at fifteen miles we encamped, for Jimmy and I were both alike hungry and thirsty. Our small supply of water only tantalized us, whenever we took a drop. On camping we found we had actually nothing to eat—at least nothing cooked, and we had to sacrifice a drop of our water to make a Johnnycake. It was late by the time we had eaten our supper, and I told Jimmy he had better go to sleep for a bit, if he felt inclined. I then went after and tied up the horses, which were rambling off. When I returned I found Jimmy had literally taken me at my word, for there he was fast asleep amongst the coals and ashes, where we had cooked our cake. I rolled him over once or twice, to prevent his catching fire, but he never awoke. The night was very warm, and no bedding required whatever. I lay down upon my rug in no very enviable frame of mind or body, for I was in pain all over from the recent kickings I had received from Diaway. I was again disappointed in discovering any feature, that would take me any further away to the west. I lay sleepless and restless, till half-past two; I saddled up all the horses and got everything ready, then I roused poor Jimmy up, much to his distress, though he was quite willing and anxious to be off. I however am getting in advance of my time, and as I did not sleep, am forgetting, that another night has passed or is passing away, and it was now morning of

Tuesday, 24th February.—Thermometer last night 88°, this morning 78°. We were in our saddles at three o'clock, and I continued on steering by the stars until daylight, travelling over sandhill after sandhill; but not generally quite so high as those traversed by us in going out, we being now a little north of our outgoing line. Soon after sunrise we fell in with our old tracks, and continued on, though we had great work to keep the horses up to the mark, until we reached the place where we had encamped the night before last. We were now only fifteen miles from the water; but the horses for the last few miles had gone so dreadfully slow, that I was afraid they were all going to give in. Directly they were unsaddled they all lay down, rolling and groaning fearfully. To see a horse in a state of great thirst is a terrible sight, the natural cavity beneath the tail opens to an extraordinary size, and the creature strains and makes the most lamentable noises. Mares are generally worse in these cases than horses. These were all in a bad state, though Diaway far less than the others. We had yet a pint or two of water in our bags, and it was absolutely necessary to sacrifice it to the horses, if I wished them ever to return. We had only three pints, which we gave to Buggs and the old mare, Diaway getting none; the others only getting sufficient just to moisten their tongues. We stayed here about an hour, leaving it at eleven a.m., and we reached the gorge at sundown, after going at the rate of only two miles an hour. The day was fearfully hot, the thermometer at eleven o'clock standing at 104° in the shade. By the time we reached the Camp-tree and took off the saddles, the horses could only stand when in motion; old Buggs could not keep his legs at all, but fell immediately in going up the gorge to water. They all fell, even the old mare, and it took nearly an hour to get them up to the bath, when they were too weak to prevent themselves from slipping in; Buggs and Diaway slipping in drinking and swimming at the same time. At last old Buggs touched the bottom with his heels, and stood up on his hind legs with his fore feet against the rock-wall and his head bent down between them, and drank thus. I never saw a horse drink in such a fashion before. It was late enough, when we got them back past the Camp-tree and let them go without hobbles. The ants were as rampant as ever, and I passed another night walking up and down the glen. Towards the middle of the night the horses came again for water; but they would not return, preferring to remain up there all night rather than risk the return passage in the dark. I went right up on the top of the mountain, and there I did get an hour's peace just before daylight.

Wednesday, 25th February.—Thermometer last night 94°, this morning 80°. The horses having remained standing round the water, their legs got all puffed from bruises in falling about the slippery rocks, and I had great trouble in getting them down again at all. I was again compelled to remain to-day in this *inferno*, as the horses were quite unable to carry us away. Dante's gelid lowest circle or City of Dis, could not cause more anguish to a forced resident within its bounds, than did this frightful place to me. My only consolation was my marble-bath, within whose cool and pellucid depths I could alone find respite from my tormentors. Oh, how earnestly did I wish its waters were the waters of oblivion, or that I could quaff some kind nepenthe, which would make me oblivious to my woes. For the persistent attacks of the ants unceasingly continued

"From night 'till morn, from morn 'till dewy eve."

but here of course we had no dewy eve. Thermometer stood at 103° in the shade to-day.

Thursday, 26th February.—Thermometer last night 92°, this morning 80°. With what pleasure I again departed from this odious gorge may well be imagined after passing another night of restlessness, misery and agony, though of course I was greatly indebted to it for the supplies of water, it had afforded us. There was a good supply of water in this reservoir, and I called it "Edith's Marble-Bath," after my niece, Miss Gill, of Kew, Melbourne, after whom I also named Glen-Edith on my former expedition. Although this was not marble, it appeared remarkably like it; there is no limestone in this range. We kept away from the roughest part of the hills more southerly than when we had come up, and not going up into any of the gorges to look for water in the middle of the day, we got back to Glen-Helen by night: having travelled thirty-three miles. Old Buggs was completely done up, and had to be led and driven on foot from the time we passed Edith Hull's Springs, three miles back, and it was very late when we got him to camp; but there being good green feed the horses did pretty well. There were, comparatively speaking, no ants here, and I had only to shift my quarters four times on account of the few (millions) that were here; I scarcely think, that a bed of bulldog ants would induce Jimmy to shift, when once down. Thermometer to-day rose to 102° in the shade.

Friday, 27th February.—Thermometer last night 86°, this morning 78°. Having only twelve miles to go to reach the Pass, we arrived there early, and found that Mr. Tietkens and Gibson were quite well, and that nothing had gone amiss while I was away. No natives had appeared during my absence; but they had seen nothing of the horses since we left.

Saturday, 28th February.—Thermometer last night 80°, this morning 68°. A fine cool night, and as here there are really no ants, I think I must have slept nearly all night, it being the first night's sleep I had for a week. There was a little drainage water in a small creek about two and a-half miles to the south of this pass. Gibson and Jimmy went out to look for the missing horses, also to try and get a shot at some emus, which used to come for water to that spot: they very soon returned, having come across all the horses. There was a marked improvement in their appearance, they have certainly thriven well. There is fine green feed here, and it is a splendid place for an explorer's depôt, being also a very agreeable and pretty spot. Gibson and Jimmy renewed their hunt for emus, but returned without any. Thermometer to-day 100° in the shade.

Sunday, 1st March.—Thermometer last night 86°, this morning 74°. We got our usual supply of pigeons last night for our morning's breakfast, but each day we can only lament, that the end of our ammunition is at hand. For dinner to-day we got some hawks, parrots and crows. I don't know which of them it was, that disagreed with me, but I had to abandon my dinner almost as soon as I ate it. I suppose the natural antipathies of these creatures to one another exist even after death and burial. It being now absolutely necessary to kill horses—as our ammunition couldn't last much longer—Mr. Tietkens and I went to find a spot to erect a smoke-house, which required the soft bank of a creek. We got a place half-a-mile away. The day was very hot, the thermometer rising to 104° in the shade.

Monday, 2nd March.—Thermometer last night 86°, this morning 72°. To-day Mr. Tietkens and I commenced operations at the smoke-house, and the first thing we did was to break the axe-handle; consequently the work was postponed for a day. Gibson, who thought he was a carpenter, blacksmith and master of many other trades by nature, volunteered to make a new handle, which no one objected to in the least. Thermometer 102° in the shade.

Tuesday, 3rd March.—Thermometer last night 86°, this morning 72°. Gibson's new axe-handle lasted until the first stick required was almost cut in two, when it came in two also; so we had to return again to camp, while Gibson made a new one on a new principle. To-day was much cooler than usual, the thermometer not rising higher than 96° in the shade.

Wednesday, 4th March.—Thermometer last night 86°, this morning 72°. Gibson made another handle and shod a horse or two, while Mr. Tietkens and I worked at the smoke-house. Thermometer 96° in the shade.

Thursday, 5th March.—Thermometer last night 76°, this morning 68°; a cool and agreeable night. Two of our horses, a pair of poking brutes, are away by themselves, so Mr. Tietkens and I went to look for them, but could not find them. We also took the shovel with us, to fill up with sand the place 2½ miles off, where the horses have been used to drink, so that they must come and water at the Pass, which soon induced them to show themselves, which they did at night. Thermometer 96° in shade.

Friday, 6th March.—Thermometer last night 72°, this morning 72°. To-day was sultry and close; thermometer 96°. Shod two more horses, and worked at the smoke-house.

Saturday, 7th March.—Thermometer last night 82°, this morning 82°. Sultry and cloudy night; this afternoon cloudy, and one slight rumble of thunder was heard; thermometer 104°.

Sunday, 8th March.—Thermometer last night 86°, this morning 68°. Cloudy night, a few drops of rain fell at one a.m. for about a minute. Day rather inclined to be cool; thermometer 94°.

Monday, 9th March.—Thermometer last night 76°, this morning 68°; clear cool night. I intended to-day to go with Mr. Tietkens to the eastward and discover a line of country for about 100 miles; indeed I wished to know, if a retreat could be made from here eastward, as it must be remembered I reached this range from Fort Mueller in the south. I intended to await a change of weather in hopes of rain, before I made another attempt to penetrate west; the sky was again beginning to get cloudy, and I had hopes of rain at the approaching equinox. The horses we required did not come in to water until night, so we hobbled and put them through to the north side of the Pass. Thermometer to-day 100° in the shade.

Tuesday, 10th March.—Thermometer last night 86°, this morning 84°; a cloudy sultry night and morning. Getting our horses pretty easily this morning, we turned eastward as soon as we got clear of the Pass, being again on the north side of this range, and intending to traverse it and others that continued beyond it to the east. Almost immediately at starting we found a pass between this range and another, running more northerly; the pass lay nearly east, we reached it in five miles. I called it "Weld-Pass," after His Excellency the Governor of the province, in which these ranges are situated. From here we had a good view of the country further east, a curved line of abrupt-faced hills traversed the northern horizon; they had a peculiar and wall-like appearance; they appeared to end at a singular looking pinnacle thirty-four or thirty-five miles away and nearly east; the abrupt-faced range to the north of us swept round in a half circle northward, and thence to the pinnacle. We kept along the Rawlinson-Range, thinking we might find some more good gorges or glens before it ended; but in this we were disappointed, for though we searched in several places, there was no water in any of them, the country being very rough and stony on the line of our travels, but our three horses were shod. We found the Rawlinson-Range ended altogether in about fifteen or sixteen miles at the Mount Russell, which I had named on my second visit to it. Other ranges rose up to the east, the most westerly or south-westerly hill of which I called "Mount Deering," after Mr. Samuel Deering, late of the Crown Lands Department. The intervening country between Rawlinson's and the new range to the east seemed filled pretty well up with scrub. Leaving the Rawlinson-Range where it ended, we pushed on for the pinnacle in the northern range; we did not reach it by night, as we stayed searching for water westward of it, and camped at twenty-nine miles without water. The day had been hot, close, sultry and occasionally cloudy; the thermometer rose to 101° in the shade, and the whole atmosphere was most oppressive.

Wednesday, 11th March.—The thermometer last night 84°, this morning 80°; the morning cloudy. Along the foot of the hills, where we now travelled, the country became scrubby, and as we approached the pinnacle, and for two or three miles round it, it was I may almost say impenetrable. From the western side of the pinnacle a small creek issued, and when we turned up it we soon found green rushes in its bed. I felt certain now of finding some rock-water at least. A little further up we saw a much brighter green patch of rushes, and amongst them a fine little pond of water. As we were so near the rocks, and feeling sure of a larger supply, we went up further, and found a splendid basin of overflowing water, which filled several smaller ones below. We could hear the sound of splashing and rushing waters, but could not see from whence the sounds proceeded. Our horses having beeen without water last night, we decided to camp here for the rest of the day, as it was such an excellent water, though we had travelled only ten miles. This morning we had passed through two fresh fires, which the natives had made at our approach. They followed us up to this water, but as they made but little demonstration, we took but little notice of them, and they went away. I suppose they were annoyed at our finding this remarkably well-hidden water. I think they imagine the instinct of our horses brings us and them to the waters we discover, and I do not think there are many waters, which we do not find out. This is a most singular little glen. There are several small mounds of stones here, placed at even distances apart, and though the ground was all stones, places like paths have been cleared between them. There was also a large piece of rock in the centre of most of these strange heaps. They were not very high, not more than two and a-half feet. I have concluded, it may be said uncharitably, that these are small kinds of teocallis, and that on the bare rock already mentioned the natives have and will again perform their horrid rites of human butchery, and that the drippings of the pellucid fountains from the rocky basins above have been echoed and re-echoed by the dripping fountains of human gore from the veins and arteries of their bound and helpless victims. The day was again warm, the thermometer rising to 102° in the shade. Here we could enjoy a cool and luxurious bath. The largest basin was by no means deep, but the water was running into and out of it with considerable force. We searched about to see where it came in by the sound it had, and eventually found, some little distance off, on the left-hand a crevice of white granite, where it came down through from the upper rocks, and ran away under our feet into rushes, and thence percolated into the large basin. On the sloping face of the white granite, and where the water ran down it, was a small indent or smooth chip exactly the size of a person's mouth, so that we instinctively put our mouths to it and drank of the pure and gushing fluid. I firmly believe this chip out of the rock has been formed by successive ages of the native population placing their mouths to and drinking at this spot, whether in connexion with any sacrificial ceremony or not, deponent sayeth and knoweth not. This water I believe is undoubtedly permanent, though it being so secluded by scrub, and there being very little good grazing country near it, it is only valuable as a wayside resting-place. I called the singular pinnacle, which points out this water to the wanderer in this wild, "Gill's Pinnacle," after my excellent brother-in-law, Mr. G. D. Gill, of Melbourne, a subscriber to the fund for my expedition; and the water I have named "Gordon's Spring," after my young nephew, Gordon Gill, his son.

Thursday, 12th March.—Thermometer last night 94°, this morning 86°. The night was hot, sultry and cloudy, and at 1 a.m. this morning the rumbling of thunder was heard. Getting our horses reasonably early for so scrubby a place, we were about starting, when some natives again made their appearance, but there were not more than half a dozen of them. They vented their displeasure at our appearance here by the emission of several howls, yells and gesticulations, and of course to frighten us to death, as they supposed, set fire to the triodia all round. We took no more notice of them, than of so many stones, but rode off and left them to their own devices, and we saw them no more. The country to the east, towards the other range in that direction, seemed very scrubby. We went first for a hill, which bore a good deal to the south of east, and crossed a broad sandy and stony creek running north, which I called the "Rebecca," after my sister, Mrs. G. D. Gill. From it we went through occasional open patches on to a saddle between two hills, all the while having a continuous range to the north, being the continuation beyond Gill's Pinnacle of the crescent-shaped range we had seen, of which that hill formed a part. Another most conspicuous mountain in the northern range, distant from Gill's Pinnacle twelve miles, I called "Mount Sargood," after the Hon. F. Sargood, Melbourne, a subscriber to the fund for my expedition; Mount Sargood lay just north of the saddle over which we went. From this saddle we now had a range on the south of us. The southern range ran up from the south-westward, and formed a valley between it and the northern range. I called the new range to the south, and which lies east of the Rawlinson, "Petermann's Range," after Dr. Augustus Petermann, of Gotha, the celebrated geographer, who has laboured so strenuously in the furtherance of all plans for Australian discovery, and who has produced the most excellent and reliable maps of this continent, which have ever appeared. In Petermann's Range a peculiar notch bore from this saddle 25° south of east. This range was exceedingly wall-like and very steep, having a serrated ridge all along. We made for this notch, hoping to find water; it proved however only a rough gully and not a pass, as I had expected. The range bore up to the north-east, and we continued along its northern foot. At four miles from the notch we came to a pass, where two high hills stood apart and allowed an extremely large creek, that is to say an extremely wide creek, to come through, whose trend was northerly. Climbing one of the hills, I found this creek came from the south-westward, and was here joined by another from the south-east; that there was an exceedingly fine and pretty piece of park-like scenery enclosed almost entirely by hills, the Petermann-Range forming a kind of huge and outside wall, which enclosed a mass of lower hills to the south, from which these two creeks, which here debouched through this pass northward, found their sources. This was a very extraordinary place. I searched in the pass in vain for water, and could not help wondering where such a watercourse could go to. I called the large creek "The Docker," after Messrs. T. B. and F. G. Docker, of Victoria, subscribers to the fund for my expedition. The pass and the park just within it I named "Livingstone-Pass," and "Learmonth-Park," after Messrs. F. and S. Livingstone Learmonth, of Ercildoun, Victoria, subscribers to the fund for my expedition. Immediately outside the pass, a little to the east of north, lay a high and abrupt-faced mountain, which I named Mount Skene, after the Hon. William Skene, M.L.C., of Victoria, a subscriber to the fund for my expedition. Our search about the pass for water was unsuccessful, and I decided to go up the more easterly of the two creeks that joined, the easterly one being very small compared to the Docker-River. By the time we had got two or three miles from the pass it became quite dusk, and we had to encamp without water, having travelled twenty-eight miles. The thermometer rose to 102° in the shade to-day. This range (Petermann's) is most singular in construction; it rises on either side almost perpendicularly, and does not appear to have very much water about it; the hills indeed seem to be mere walls.

Friday, 13th March.—Thermometer last night 90°, this morning 80°; at two a.m. this morning we heard the slight rumblings of thunder and saw one or two flashes of lighting, which cooled the air a little; there was very fine grass here, and the horses stayed well. At starting we continued on up this creek, which appeared to rise in some low hills to the south, though it wound about so much, that it was only by travelling we found it went more easterly towards a peculiar ridge, upon whose top was a fanciful looking broken wall or rampart with a little pinnacle on one side; the creek came from the low hills near the foot of those higher ones, and when nearly abreast of the little pinnacle we found some water in the creek-bed, which now became very stony; the water was not very good, but the horses drank it with avidity. Above this we soon got some very good water in rocks and sand; I called this queer-looking wall "The Ruined Rampart." There was a large quantity of greenish and ammonia-tasting water in some of the rock-holes here, some saltish, others nearly putrid; one or two ducks flew up from these strange ponds. We remained here a couple of hours. There was an overhanging ledge or cave, which gave us good shade, the morning being very hot. We left this singular place, where I believe a good supply of water is always to be obtained, at twelve o'clock, first calling it "MacBain's Springs," after James MacBain, Esq., M.L.A., of Toorak, a subscriber to the fund for my expedition. When we left, the thermometer stood at 102°; in the afternoon it was much hotter. On leaving here we continued to follow up the creek; it took its rise in a mass of broken tablelands, we still having the high walls of the Petermann-Range to the north and very close to us. Our course was now a few degrees to the south of east; in five miles we got over the water-shed and descended the rough bed of another creek running to the eastward; it also had some very queer water in it, being pink, green and blue; there were a few ducks here also, and some good water was obtained by scratching in the sand. I called these "Harry Hull's Springs," after my sister's step-son, Mr. Harry Hull, jun., of St. Kilda; and the creek I named after him also. Following it down it traversed a fine piece of open grassy country, which was very pretty and excellent travelling ground. This was a very park-like piece of scenery. This creek joined another, which we struck in a short distance. The new creek was of enormous size, as far as width is concerned; it came from the lower hills to the south, and ran north, where the main ranges parted to allow its passage. The natives were burning the country some distance down the creek. Where on earth can it go to? No doubt water exists in plenty up at its head, but there was none where we struck it. I called this "The Hull," after the Hon. Wm. Hull, of St. Kilda, my brother-in-law's father, and a subscriber to the fund for my expedition. The main range now ran on in more disconneced portions than formerly, whose general direction was 25° south of east; we still had a low mass of hills to the south, but they were some miles away. We continued on, travelling under the lea of the main range until night, when we camped without water, having travelled 25 miles from the Ruined Rampart. A few miles before we came to camp we passed a high cone in an isolated range, which formed the eastern part of a valley or pass through which Hull's Creek runs to the north; I called this "Mount Curdie," and the range of which it is the highest point, "Curdie's Range," after Dr. Daniel Curdie, M.A., of Camperdown, a subscriber to the fund for my expedition.

Saturday, 14th March.—Thermometer last night 82°, this morning 70°; a cool night and morning. Before starting this morning I ascended a hill under which we had camped, being the most eastern point of Curdie's Range, while Mr. Tietkens went for the horses. From here I could see the top of a peak bearing 27° south of east, a long way off over the tops of other hills—this I supposed to be (as it ought) what I had called Gosse's Sugar-Loaf on my outgoing journey from Mount Olga, and what Mr. Gosse called Stevenson's Peak. The view to the north from here was bounded by a long wall-like line, stretching across the horizon, but ending about north-east; they were mostly disconnected chains and apparently of the same character as the Petermann-Range. This I named "Blood's Range," after Mr. Blood, of the Peake. I could trace the Hull-Creek for many miles winding away to the north and a trifle west of that point; it is evident, that there must exist some gigantic basin, into which the Docker and the Hull must flow. These creeks appear to rise in the mass of low ranges, which lie southward; and as the line of them continues (which I can see from here to be the case) it is highly probable, that I shall meet yet more creeks running in the same direction. I feel morally sure in my own mind, that the Lake Amadeus of my former journey is the basin, into which these creeks descend; and if there are creeks running into it from the south, may there not also be others running into it from the north and west? The line of the southern hills, of which I have just been speaking, now runs across the bearing of Gosse's Sugar-Loaf; so I shall have to pass over or through them to reach that hill, and as I said before the bearing of it was 27° south of east. The main or outer-line of the hills, on which I am now, still runs on in disconnected groups in nearly that same direction, forming a kind of back-wall all the way. Making a start, we came at seven miles to where the first hills of the southern ranges approached our line; they were mostly disjointed, having small grassy valleys between them. They were here mostly festooned with cypress-pines, and presented many huge bare rocks, being very similar country to that described by me at Ayers-Range, through which I passed last August. At ten miles we reached a small creek with eucalypts, and obtained water by digging; it was rather saltish, but our horses were very glad of it; we gave them a couple of hours' rest. I called this "Louisa's Creek," after my sister's step-daughter, Miss Louisa Hull, of St. Kilda. From here a mount, standing by itself in the northern line of range, and being immediately eastward of Curdie's Range, bore 13° north of west from this water; I called this hill "Mount Fagan," after my old and valued friend and former brother official Mr. T. B. Fagan, of the Post Office, Melbourne. Another hill, eastward again of Mount Fagan, bore from this water 42° east of north, I named this hill "Mount Miller," after my very kind friend, Mr. Miller, of Anlaby Station, who entertained Mr. Tietkens and myself most kindly there, when we were going out on this expedition. At five miles from here we struck another and much larger creek, running up to the north; and close to where we crossed on our right hand, was a rocky gorge or fall, over and through which the water tumbles in times of flood. There was plenty of water up there, I do not mean to say for stock; but we did not stay here sufficiently long for ascertaining this; I called this creek the "Chirnside," after Andrew Chirnside, Esq., near Geelong, a subscriber to the fund for the expedition. Another mount, eastward of Mount Miller, but some miles away, bore from this gorge 14° north of east; I gave to this the name: "Mount Bowley," after Mr. Bowley, of the Telegraph-Station, at Charlotte-Waters, who was most hospitable to me on my short visit there, previous to my departure on this expedition. At ten miles from Louisa's Creek we camped at another creek larger than the Chirnside, but we got no water at it, as it was a long way southward, up to where it issued from the hills. No doubt there is plenty of water in it towards its source; it ran up to the north. I called this "The Shaw," after Thomas Shaw, Esq., of Camperdown, a subscriber to the expedition-fund. We crossed over one or two other creeks also, which ran north, but no doubt joined the Chirnside or the Shaw in a few miles. We were now in the midst of broken pine-clad hilly country, very well grassed and very pretty, having the high hills I just named on the north, and low hills on the south; the day was moderately cool, the thermometer not rising higher than 98° in the shade. Ever since we entered the Livingstone-Pass yesterday, we have traversed country, on which the odious triodia has been remarkably absent, but there was a little at the entrance to the Livingstone-Pass.

Sunday, 15th March.—Thermometer last night 76°, this morning 64°; a fine cool night. Travelling along this morning through this "wild Parthenius tossing in waves of pine" at six miles along our course, which is now, and has been, towards Stevenson's-Peak—for the configuration and trend of the whole lines of hills both north and south of my line of march runs in that direction—we came to a place, where we surprised some natives hunting. They however almost instantly became aware of our presence from their acute perceptions of sound, scent or sight, and as is usual with these persons, the most frantic yells rent the air. Signal fires were lighted immediately, in order to collect the whole tribe, and before we had gone more than a mile, we were pursued by a multitude of howling demons, who surrounded us in all directions, though the greater bulk of them followed after us, but all yelling, screeching and howling, and rattling their spears, and running up behind us with the evident intention of never allowing us to depart out of their coasts. At one place they drew up in a body and so closely surrounded us, that we were compelled to get our revolvers ready for immediate use. We had no gun or rifle with us. We were compelled to stop and turn round, to face them and let go the pack-horse. A number of them from behind sent a lot of spears. One just grazed the pack-horse's hide and made him run about. Just as we charged at them and fired off our revolvers, I was horrified to find, that my pistol would not go off, something being wrong with the cartridges; for though I snapped it four times not a single discharge took place. Fortunately Mr. Tietkens's went off, and what with that and the pack-horse rushing about loose, it drove the wretches off, for a time at least. They seemed infinitely more alarmed at the horses, than anything else. They certainly had no fear whatever of us. At all events they retired for a bit, and we proceeded after breaking about forty of their spears. The country had now become much rougher, the little grassy valleys having ceased, and we had to take to the hills. Occasionally we got a glimpse of the peak, for which we were steering. At nine miles from where we had confronted our enemies we came to a bold, bare, rounded hill, ascending which we saw immediately below us this peculiar hilly country ceasing to the east, though the same kind of hills ran on south-easterly. Two or three small creeks were visible below, then a thick scrubby country set in, bounded exactly to the east by Mount Olga itself, which was sixty miles away. The peak we had been steering for, and which proved to be the one I thought it was, bore 30° south of east. There was much bare rock all round this hill, and in one of the crevices we got a little water; we turned our horses out here. While indulging in our dinner, Mr. Tietkens gave the alarm, that the natives were upon us again, and instantly we again heard their discordant yells; the horses commenced to gallop off in hobbles. These wretches now seemed determined, to do us some grievous bodily harm, as they came and surrounded us again, their numbers considerably augmented. Now there were sixty or seventy of them; two of these men were of commanding stature—one more especially so—large enough to make two of Tietkens, if not of me. We did not find these giants the most forward in the onslaught; probably they were only brought out for show. The horses having gone a good step, Mr. Tietkens went after them, while I, in some trepidation at the thought of my revolver playing me false (of course I had reloaded it), prepared to defend the camp. They immediately swarmed round me, running until they came within fifty yards, then coming slower. I had no alternative but to face them, and, as they were coming towards me, I walked very fast towards them, and presented my revolver at them. Some of them, thinking, I suppose, I was only pointing my finger at them, pointed their fingers at me, while nearly all the others had their spears ready to discharge; and I am sure that I should, in another instant, have been transfixed with twenty or thirty spears, had not Mr. Tietkens tied up the horses, and come running up also; and both our revolvers going off properly enough this time, we made our foes retreat at a better pace than they advanced. A great many dropped their spears and other weapons, and went scudding away over the rocks, as fast as fear and astonishment would permit. We broke all the spears, we could lay our hands on, numbering nearly a hundred, and then finished our dinner. After dinner we found a better supply of water, in another creek about two miles southward; there was both rock- and sand-water there. Water, I believe, is always obtainable by digging at all the watering-places, lately discovered by us. We had come a considerable distance now from Sladen-Water, and found waters all the way. Mount Olga was in sight; but the question was, is the water there permanent? There is no digging to be done there—it is all solid rock; you either get it on the surface, or there is none. My readers must believe, that I made this trip to the east not with any desire to retreat or to return; but it was necessary to wait for a change of weather, before I could hope to be successful in penetrating to the west; therefore, to occupy my time, and to become acquainted with as much country as I possibly could, I had come to the east. We had come over 140 miles, from our camp at the Pass. The sight of Mount Olga, and the thoughts of returning to the eastward, acted as a spur to drive me further to the west; therefore we again turned our backs upon Mount Olga and the distant east. I named this gorge, where we found the good supply of water, "Glen-Robertson," and the creek in which it exists, "Casterton-Creek," after George Robertson, Esq., of Casterton, a subscriber to the fund for my expedition. Mount Olga, as I said before, bore east from here; its appearance from the west was most grotesque; it seemed like five or six enormous pink haystacks, all leaning up against one another, with crevices between, which came only about half-way down its face. I should think this is one of the most extraordinary mountains on the face of the earth; for, as I before mentioned, it is composed of several enormous rounded stone-shapes, like the backs of several monstrous kneeling pink elephants; and at sixty miles to the west from it its outline is astonishing; the highest point of all, looking at it from here, presents the appearance of a gigantic pink damper or Chinese gong, viewed edgeways, and slightly out of the perpendicular. Not returning to where we had dined and fought, we went about two miles northerly. We had to take to the rough hills again, or go a long way further north. The hills were exceedingly rough; we had to wind in and out the best way we could for three or four miles, when we struck our out-going tracks. It was rather difficult to follow them through these stony hills. We found, that the natives had followed us up on them step by step, and had tried to stamp the marks of the horses' hoofs out of the ground with their own feet, which made it much easier for us to follow, and consequently *re*-mark. They had gone four and five abreast, and endeavoured permanently to efface them. We saw them raising puffs of smoke behind us, but did not anticipate any more annoyance from them. We pushed on until dark, when we reached the place, where my revolver would not go off in the morning, and there we encamped without water. I omitted to mention, that after encountering the natives here, I first heard and then saw one of those very large iguanas, which are supposed to exist in most parts of this country. I have heard great tales of their size and ferocity. The natives near the Peake, I believe, call them

"Parentys." The one seen to-day was nearly black, and from head to tail over five feet long, being exactly similar to those, found near Mount Corong and elsewhere in Victoria, and in the stringybark forests of New South Wales.

Monday, 16th March.—Thermometer last night 74°, this morning 66°. Before daylight this morning I went for the horses, while Mr. Tietkens got the breakfast ready, and fixed up the swags. We had only just sat down to our meal, when Mr. Tietkens, whose nervous system seems particularly alive to any native approach, gave the alarm, that our pursuers were again upon us, and immediately afterwards we were again saluted with their hideous outcries. Breakfast was now a matter of minor import, than we had previously considered it. By the time we got the saddles and things on the horses, we were again surrounded. How I wished we had only one of our Snider rifles; but unfortunately we had left them at home. We could do nothing with such an insensate insatiable mob of wretches as these; so we got into our saddles and fairly gave our enemies the slip through the speed of our horses; they ran a long distance, but the pace told on them at last, and we completely distanced them. Had we been unsuccessful in finding water in this country, and then have met these demons, it is more than probable, that we should never have escaped. I have encountered natives often enough in most parts of the country, but these were the most persistent persecutors I ever met, and I don't sigh to meet them again; the great wonder is, that they did not sneak upon us in the night; but nocturnal perambulation does not appear to be a habit of the natives in this part of the Continent. We kept at a good pace until we reached the Chirnside, and gave our horses a drink; but went on to Louisa's Creek before we rested, having come twenty miles. We only remained here an hour. We saw no more of the natives; but we still pushed on twenty-two miles further, until we reached the Hull, where we encamped without water. The thermometer had not risen above 100°. Our supply of food now consisted of just sufficient flour, to make two small Johnny-cakes: and as we had still over eighty miles to travel, we simply had to go without any food all day, and shall have precisely the same quantity to-morrow—that is to say, *nil.*

Tuesday, 17th March.—Thermometer last night 74°, this morning 66°. At eleven miles we reached the water near the cave under the Ruined Rampart, and which bears south 15° west from the little pinnacle, which forms part of the ruin, and which I named MacBain's Springs in coming out. Giving the horses an hour or two here, we again pushed on and travelled twenty-three miles by night on our old tracks, and camped without water for the horses or food for ourselves. The thermometer had stood at 100° in the shade of the cave, which we left at twelve o'clock.

Wednesday, 18th March.—Thermometer last night 74°, this morning 78°. At thirteen miles we got back to Gill's Pinnacle; here while almost starving we mixed up our small remnant of flour. The thermometer to-day rose to 98° in the shade. While we remained at the Pinnacle, we found that during our absence from this place the natives had placed a quantity of eucalyptus-leaves or small boughs into the interstices of the small mounds of stone, or (as I called them) teocallis, which I mentioned previously; they were evidently put there as soon as we departed, for they were now dead and dry. Remounting our horses we made good another twenty miles by night, and camped in casuarina- and triodia-sandhills.

Thursday, 19th March.—Thermometer last night 82°, this morning 72°. A warm sultry night. We reached the camp at the Pass by 9 a.m. Gibson and Jimmy were there and nothing had gone wrong in our absence; these two poor fellows looked as pale as ghosts, when we returned. Gibson imagined we had gone to the westward, and was in a great state of mind at what he considered our protracted absence. The water in the water-hole did not agree with him, and at starting I had given him half-a-bottle of rum, the last of our medical stores. I had also told him, where to look for a spring of good water close to the camp, and where I had got bogged one day; but he had not made any attempt to find or open it out. The thermometer to-day stood at 100° in the shade.

Friday, 20th March.—Thermometer last night 84°, this morning 70°. Last night we had thunder and lightning and a few drops of rain. We were now completely without any food in the shape of meat, and had but a very limited quantity of flour. Gibson and Jimmy had shot plenty of pigeons and parrots and one or two ducks, while we were away, but the ammunition was now so greatly reduced, that a single shot is of the greatest consideration. I said formerly, that one horse was fattening for the knife; Gibson reported that he now looked very well and fit for market; so I determined to kill him as soon as the smoke-house was thoroughly complete, and Mr. Tietkens and Jimmy went to-day to put a few finishing strokes to it, while Gibson and I cut up tarpaulins to make some pairs of large water-bags, and with a small lot of new canvas we made four pairs large enough to hold seven or eight gallons each; these I considered, when well greased with horse fat (or oil), ought to enable me to get some distance from the western extremity of this range. Old Billy or Terrible Billy, as we always called him, came to water early, and I was much pleased with his appearance; but I determined to complete the water-bags before killing him; so he has a day or two's longer grace. The day was cloudy and cool. Thermometer only rose to 88°. At night we had violent squalls of wind, and about two minutes' heavy rain.

Saturday, 21st March.—Thermometer last night 72°, this morning 70°. This day of the autumnal equinox I had looked forward to eagerly in hopes of rain falling. The morning was again cloudy. In the course of the day we had two or three dry thunderstorms, and rains seemed falling in all directions, but only a few drops visited us *en route* to some more favoured land. The day was cool; Gibson and I hard at work at the water-bags; thermometer 80° in the shade.

Sunday, 22nd March.—Thermometer last night 76°, this morning 62°. To-day was actually a day of rest, one of the few Sundays I have been able to call so since I started. The day fine and cool, thermometer not rising above 80° in the shade. Near this camp there were several strange little heaps of stones or rather, as a general rule, small circles, where the stones had been removed with the exception of a solitary one left in the middle of the circle; for what purpose the natives could have made or cleared these places I cannot imagine.

Monday, 23rd March.—Thermometer last night 70°, this morning 60°. A fine cool night. We are still at work at the water-bags. Thermometer rose only to 78° in the shade to-day.

Tuesday, 24th March.—Thermometer last night 72°, this morning 62°. Finished the water-bags at last. Thermometer 80° in the shade. The weather for the last few days has been truly delightful.

Wednesday, 25th March.—Thermometer last night 74°, this morning 64°. There were still some sticks to be got for the smoke-house, which Mr. Tietkens and I got, while Gibson took the shovel, to open out the springs formerly mentioned; they lie in the midst of several little clumps of young eucalyptus-suckers, the ground all round being a morass, in which a man might almost sink, were it not for the thick growth of rushes. The water appears to flow over several acres of ground, but appearing and disappearing in places. The moment a small space was cleared of the rushes it was evident, that the water was perpetually flowing, and we stood over our ankles in black soil. Gibson dug a small tank, and the water soon cleared for itself a beautiful little crystal-basin of the purest liquid, so much more delicious and wholesome than the half-brackish water in the bed of the creek. These springs have their origin at the foot of the hill on the eastern side of this pass, are permanently flowing, and percolate into the creek-bed, where the water becomes impregnated with salt or soda. The water in the holes is always running. I thought the supply came from up the creek; now however I find that it comes from these freshwater-springs. Any future traveller, who may visit this remote and lonely pass, will easily find the freshwater-springs, as they are not more than thirty or forty yards from the creek-bank, being on the east side and near the foot of the eastern hill, and not more than a quarter of a mile above a tree marked "E. Giles" with date.

Thursday, 26th March.—Thermometer last night 68°, this morning 50°. The old and doomed Terrible Billy confidingly came to water at eleven o'clock last night, but having set a watch for him, escape was hopeless. He took his last drink, and was led a captive to the camp, where we tied him up all night. The old creature looked remarkably well, and when tied up close to the smoke-house—innocent, unsuspecting animal! he commenced to eat a bunch or two of grass, when a Snider-bullet crashing through his forehead terminated his existence. There was some little fat about him; it took some time to cut up the meat into small pieces, which were strung on to sticks, and placed in the smoke-house; but we could not get the work quite finished by night. We made a fine supper of horse-steaks, which we now relished amazingly. He tasted much better than the cob, which we had eaten at Elder's Creek. Those of my readers, who are not aware of the internal construction of a horse, may learn, that there is neither caul nor kidney-fat; what fat there is is very yellow, and much softer than that of beef—indeed it will scarcely harden or coagulate at all. The fat mostly lies on the inside lining of the stomach, and there is some in the intestines; and with very fat horses a salvage of fat may be got on portions of the meat. The kidneys are not like those of cattle, but more like very large sheep's kidneys; they are not eatable, at least under ordinary circumstances. The thermometer to-day rose only to 80° in the shade.

Friday, 27th March.—Thermometer last night 76°, this morning 68°. The smoke-house is now the object of our solicitude. The smoke ascends from the immolated Billy day and night. The smoke-house is some little distance from the camp, as we could not find a suitable bank nearer than half-a-mile. Our continued column of smoke induced some natives, to make their appearance to-day, but they kept at a very respectful distance, coming no nearer than the summit of the hills on either side of the Pass, from whence they had a good bird's-eye view of our proceedings. They saluted us with a few cheers (*i.e.* groans) as they watched us from their elevation. The weather is now beautifully fine and clear, with light breezes from the south.

Saturday, 28th March.—Thermometer last night 72°, this morning 62°. Smoking "Billy" and eating "Billy" all day. Wind south; fine, cool day. Thermometer 82° in shade.

Sunday, 29th March.—Thermometer last night 72°, this morning 62°. Smoking "Billy." Wind north; day fine and cool. Thermometer 80° in shade.

Monday, 30th March.—Thermometer last night 76°, this morning 68°. The day was hazy, though cool. Thermometer 82° in shade.

Tuesday, 31st March.—Thermometer last night 76°, this morning 72°. The day cloudy and sultry. Thermometer 86°.

Wednesday, 1st April.—Thermometer last night 76°, this morning 74°. The day hot, cloudy, and sultry. Thermometer rising to 94° in shade.

Thursday, 2nd April.—Thermometer last night 76°, this morning 70°. Morning cloudy, but clear towards afternoon. Thermometer 90° in shade.

Friday, 3rd April.—Thermometer last night 72°, this morning 64°. Windy night; wind from east. I mentioned before, I intended to make further attempts to push west from the end of this range. We had now finished smoking Terrible Billy, who still maintained his name, for he was terribly tough. All we now required was, to get the horses and again start away, but that was not the easiest matter in the world, for they were all running loose, and though they had to come for water to the Pass, yet they mostly came by twos and threes, and that generally in the night. This morning before daylight I heard some at the water, and we managed to get a couple and tied them up till daylight. Mr. Tietkens and Gibson went to look for the others, but they only found ten, making twelve in all. These we short-hobbled, and put down through the Pass to the north; and as I now intended to keep watch for the others they couldn't well get by, without the person on watch seeing or hearing them, as it was fine moonlight most of the night. Thermometer 86° in the shade.

Saturday, 4th April.—Thermometer last night 78°, this morning 66°. Only two horses came last night to water, but they got away before we could catch them through some blunder in letting them drink before catching. No more came during the day. Thermometer 88° in shade.

Sunday, 5th April.—Thermometer last night 76°, this morning 66°. Late last night two more horses came, and were put up with the others. To-night we kept a horse saddled to be sure of keeping what we had, and catching the others as they came. During my watch (the second) four horses tried to pass the camp, and I drove them back twice, and had no more trouble with them; but this morning, when we went to muster them, every hoof was gone. Of course, nobody let them go. All the others of the party informed me, that they were individually ready to take their dying oaths, that the horses never got away in their watches, and that they never had any trouble in driving them back at all; so I could only conclude that they got away in mine, as I was not prepared to make so rash an affirmation. Soon after breakfast Mr. Tietkens went away to see, if he could recover them, but soon returned, informing me that he had met a number of natives at the smoke-house, who appeared very peaceably inclined, and who were on their road down through the Pass. This was rather unusual, as since our first hostile encounter they had mostly given us a wide berth, and preferred to be out of the reach of our Sniders than otherwise. They now soon appeared, though they kept away on the east side of the creek, some little way off. They then cooeyed, when I replied and beckoned them to approach, but they sat down in a row instead. There were thirteen of them. Mr. Tietkens and I went over to them, and we had quite a friendly conference. Their leader was an individual of a very uncertain age—he might have been forty, or he might have been eighty (in the shade). His head was nearly bald on the crown, but plenty of long grizzly locks depended below the bald patch. The others were all much younger, though some of them, though not clean past their youth, yet had about them some smacks of the saltness of age. The old man was the most self-possessed, the others displaying a nervous tremor at our approach; those nearest sidling closer to their more remote (and, as they no doubt thought, fortunate) fellows. They were all extremely ill-favoured in physique, though their figures were not so *outré*, except that they appeared emaciated and starved, otherwise they would have been men of good bulk, for their legs were straight and their heights would average 5 ft. 9 in., all being much taller than Mr. Tietkens or I. Two remained at a distance. These had a great charge to superintend, being no less than that of the trained wild dogs, belonging to the tribe. There were three, and very large dogs they were, two of a light sandy and one of a kind of German colley colour. These natives were armed with an enormous quantity of spears, each having about a dozen. They do not appear to use the boomerang in this part of the continent, though we have occasionally picked up portions of old ones in our travels. Mr. Tietkens gave these natives a small piece of sugar each, with which they seemed perfectly charmed, and in consequence patted the seat of their intellectual (that is to say, digestive) organs with great gust, as the saccharine morsels liquified in their mouths. They seemed highly pleased at the antics and general appearance of my little dog, who both stood and sat up at command in the midst of them. They kept their own dogs away, I presume for fear that we might seize them to eat. The old gentleman was much gratified with my watch. I then showed them some matches, the ignition of which appeared rather too startling a phenomenon for their weak minds, and some of them rose to depart, but the old man reassured them. I presented him with several matches, and showed him how to use them, and he was much pleased; and, having no pockets in his coat, he stuck them in his hair. Mr. Tietkens during this time was smoking, and the sight of smoke issuing from his mouth seemed rather to disturb even the old man's imperturbability, and he kept much closer to me in consequence. I next showed them a revolver, and tried to explain to them the manner of using it, and they most of them repeated the word *bang*, when I used it; but when I fired it off, they were too agitated to take much notice of its effect on the bark of a tree, which might otherwise both have pointed a moral and adorned a tale in the oral traditions of their race for ever. At the report of the pistol all rose and seemed anxious to go; but I would not allow my dear old friend to depart without a few last friendly expressions. One of these natives was pitted with small-pox. They seemed to wish to know, where we were going, and when I intimated west, and by shaking my fingers to intimate a long way, many of them pulled their beards and pointed to us, and the old man gave mine a slight pull and pointed west, which I considered as significant, that they were aware, that other white people like us lived in that direction. The conference ended, and they departed over the hills on the east side of the Pass, but it was two good hours before they disappeared. In the meantime Gibson and Jimmy had caught the two horses which gave us the slip on the previous night, and which had again come to water. Thermometer to-day 90°.

Monday, 6th April.—Thermometer last night 78°, this morning 68°. All the hobbled horses which had escaped in the night now returned to water, and were again put down through the Pass; and during the day the others, which were still absent, arrived also and were secured, and all were together at last. Thermometer 90° in shade.

Tuesday, 7th April.—Thermometer last night 78°, this morning 66°. The horses were all right this morning, and we again left this delectable pass—again *en route* for the west; in leaving a long-settled camp there are always some delays, and it was noon before we finally departed, hoping to see Sladen-Water and the Pass of the Abencerrages no more. The morning was very warm; thermometer at 90° when we left. We got only fourteen miles, but were delayed by Banks (carrying my boxes), as an old strap broke, and he set to work to free himself of everything; one box fortunately with the instruments and quicksilver remaining firm. Everything got kicked out of the other one, buckskin-gloves, matches, rifle-cartridges, bottles of medicine, eye-water, socks and specimens of minerals and plants, all thrown about in the thick triodia. It was wonderful, how many things we recovered!—though some were lost, and I do not know what they were. By this time, it being dusk, the atmosphere had cooled greatly, and there was a great predisposition for a cold night.

Wednesday, 8th April.—Thermometer last night 68°, this morning 44°; the night was remarkably cold, but instantly at the sun's approach the temperature became warm enough. I now intended to camp on the range, at the fine spring called Fort McKellar, being four miles east of the Gorge of Tarns. There was a fine and heavy clump of eucalyptus-trees there, and a very convenient and open sheet of water for the use of the camp. I had always looked upon this as a very excellent and desirable encampment, though we had never used it previously. We reached it early in the day. The grass, however, is not too good or abundant, the country around being so stony and sterile. It was cloudy and sultry this afternoon; the thermometer standing at 86° in the shade.

Thursday, 9th April.—Thermometer last night 82°, this morning 74°; a sultry night, but a cool breeze from east sprung up by morning, though the afternoon again became hot and sultry. Thermometer 90° in shade. We remained here to-day, and fixed up a bit of palisade.

Friday, 10th April.—Thermometer last night 84°, this morning 74°; another sultry night. My readers will remember, that near the western termination of this range, in going along it previously—that is to say, on the 29th January—I had found a creek with a large rocky tarn of water, which we called "The Circus." I was now of course anxious to know, how the supply there was holding out; it was twenty miles west from here. Gibson and I therefore started up the range to inspect it, as it would be from there, that in any attempts to the west the last water would be got. On our road we revisited the Gorge of Tarns; the water had slightly shrunk since we left. We had left here some old articles, such as three pack-saddle frames, two pieces of trace-chain, which we had carried from Mr. Gosse's depôt near Fort Mueller, and part of a broken thermometer, and sundry other old gear, everyone of which the natives had taken away. I had a good swim in the old tarn, and we proceeded on, reaching the Circus early in the evening. The water here had become greatly reduced; it was ten weeks and two days, since I was here; in another fortnight the water would all be gone. If I intend doing anything towards the west, it must be done at once, or it would be too late. The day was warm, the thermometer standing at 100° in the shade. A large flock of galars and hundreds of pigeons came to water at night, but having no ammunition, we did not bring a gun; the water was too low in the waterhole for the horses to reach, and they had to be watered with a canvas-bucket.

Saturday, 11th April.—Thermometer last night 84°, this morning 74°. I said previously, that at the western extremity of this range lay the bed of an ancient lake. I had only been a mile or two upon its surface. There were some faint indications of salt, and I thought, further on it we might manage to get a small quantity, as we were now quite without that commodity. So we went to pay the Lake a visit on that account. On reaching it, however, no salt whatever could be got, and I had to be satisfied with getting a lot of samphire and saltbush leaves, which we took back home with us by night. Thermometer to-day 98° in the shade. I called this singular and extensive feature "Lake Christopher," after my very kind friend and namesake, Mr. Christopher Giles, jun., of the Charlotte-Waters Telegraph-Station, who most hospitably entertained me and my companions both on my departure and return from this expedition.

Sunday, 12th April.—Thermometer last night 84°, this morning 74°. Hot and sultry night. I was in hopes this oppressive weather might turn to rain, and I delayed entering that evidently frightful bed of sand, which lay to the west, in hopes of at least a change. For I must admit I dreaded to attempt this country, while the weather was still so hot and oppressive; for though, as I have before remarked, the thermometer may not appear to rise extraordinarily high in this region, yet the weight and pressure of the atmosphere is almost overpowering. I have for years traversed the driest portions of the Darling back-country, where the thermometer rises many degrees above what I have found it here. I never was much inconvenienced by the greater heats of that region; but here existence is in a perfect state of languor, and I am certain that the other members of my party feel it much more oppressive than I do. Of course not having been well fed, may make some reason for our feelings on the subject. The horses also are affected with extreme languor here as well as the men. The thermometer stood to-day at 99° in the shade.

Monday, 13th April.—Thermometer last night 88°, this morning 84°.

Hot and sultry all night, and hot and cloudy all day. Thermometer 100° in the shade.

Tuesday, 14th April.—Thermometer last night 88°, this morning 80°. Night again hot and sultry, during the day a few clouds gathered. Thermometer 98° in the shade.

Wednesday, 15th April.—Thermometer last night 88°, this morning 80°. Hot and sultry again all night, with warm wind from the north. Clouds all gone, and day very hot; thermometer 100° in the shade. The horses seemed inclined to roam away back to Sladen-Water. Mr. Tietkens and I had a very long walk after them to-day.

Thursday, 16th April.—Thermometer last night 88°, this morning 80°. Another hot night; a cool breeze sprang up east about sunrise. I was really getting anxious about the water at the Circus. I scarcely dare to grapple with that western desert in such weather; yet if I did not, I should lose the Circus-water. Although we were near the change of the moon, I despaired of a change of weather. I did not ask for rain, I only wanted the atmosphere to become a little less oppressive. I had not been round the extreme western end of this range, though we had been to it, and I thought it probable some creek might yet contain a good rock-hole on the opposite side to the Circus, and perhaps as far, if not further to the west. I therefore requested Mr. Tietkens to go with Gibson or Jimmy (the former volunteering), and try, what he could discover, and then return by the Circus, so as to report upon it also. They started away, and Jimmy and I remained at the depôt and erected some more woodwork—that is to say, rails for the fort. Thermometer to-day 98°.

Friday, 17th April.—Thermometer last night 86°, this morning 78°. Having but little to do, Jimmy and I walked away to reinspect two glens and springs lying within a couple of miles to the east of us, the first one being no more than three-quarters of a mile off, and which I now named Tyndall's Springs, as mentioned previously; it has a fine stream of running water, which descended much further down the channel from the range, than that of any other creek in the range, though it spreads into no sheets of open water as at our depôt. There is over a mile of running water here, and the channel is thickly set with fine tall bulrushes; there is also a fine shady clump of eucalyptus here, close up at the foot of the range. The next one, about a mile further east, and which I had called Groener's Springs, had not such a supply of water, though the eucalypts in the clump (for at all these waters these may be said to exist only in clumps at the extreme head of the springs) were much thicker and more numerous, and many trees, as was the case at our Fort, were of considerable size. After satisfying ourselves with our walk here, we returned home. Late at night Mr. Tietkens and Gibson came back, and reported, that, though they had discovered a new rock-hole of some size, with seven or eight feet of water in it, it was utterly and hopelessly useless; as no horse could get within three-quarters of a mile of it, and they had been quite unable to water their horses there, but had to come round to the Circus to do so. The day was warm enough, the thermometer standing at 98° in the shade. They said the Circus-water was holding out well, though it had evidently diminished since Gibson and I saw it last.

Saturday, 18th April.—Thermometer last night 88°, this morning 76°,

We remained at the Fort to-day; thermometer 96°. I got Jimmy Andrews to climb a large tree, and make a scaffold to mark it. It is branded $\frac{\text{E.G.}}{\text{4-74}}$, in large letters, on the bark.

Sunday, 19th April.—Thermometer last night 86°, this morning 76°. At the Fort all day. By evening I had made up my mind, to make some attempt at once to penetrate the sand-hills to the west—at all events in the first place to try what impression 100 miles would make on it; this I intended only as a preliminary excursion, as I had a matured plan of action to follow out, when a change occurred in the weather. I had waited and waited for a change—not to say rain, and it seemed as far off, as though the month were November instead of April. I might still keep on waiting until every ounce of our now most limited supply of rations was gone; for we were now, and had been since Billy was killed, living entirely without flour or sugar—the latter was all gone; what flour we had we kept by us for emergencies or sickness—and lived entirely on smoked horse. I had been disappointed at the Charlotte-Waters by not being able to recover a horse which I had left there, and thus started from the Alberga deficient of the two hundred pounds of flour he would have carried, which had considerably foreshortened my anticipated original supply. An enormous quantity of flour, comparatively speaking—I dare say sufficient to have made five or six good dampers—was also lost by the continual rippings of bags in the scrubs further south, and general loss of weight; but we had supplemented our supplies here and at Fort Mueller with pigeons and wallabies, as long as our ammunition lasted. I made it known to-night, that I intended to start away for the west to-morrow, so as to utilise the Circus-water yet remaining, whereupon Gibson immediately volunteered to accompany me, and complained of having previously been left so long and so often in the camp. I much rather desired Mr. Tietkens, as I knew the task we were about to undertake was no ordinary one; and I knew Mr. Tietkens was to be depended upon to the last under any circumstances; and though I said nothing, I was not altogether pleased. Mr. Tietkens waived his right to accompany me in consequence of Gibson's remarks, and he agreed to remain with Jimmy Andrews at the Fort. The day was a trifle cooler than yesterday, the thermometer not rising higher than 92° in the shade.

Monday, 20th April.—Thermometer last night 80°, this morning 76°. Gibson and I got all our things ready this morning, and taking two horses to carry water and two to ride, took with us a week's supply of smoked horse, nothing else, as we expected to be back by that time. We camped at the Circus, twenty miles west along the range from Fort McKellar. We brought four excellent horses, all in fine condition. I rode the Fair Maid of Perth, an excellent walker; I gave Gibson Badger, the big ambling horse. I packed the big cob—a splendid bay horse and fine weight-carrier—with a pair of water-bags, which contained at starting twenty gallons; the other horse was Darkie—a fine, strong, nuggety harness-horse, who carried two kegs of five gallons each, and our stock of horse-meat, rugs, &c. We reached the Circus early, and the horses had time to feed and fill themselves after being watered, though the grass was only so-so. Thermometer to-day 95° in the shade.

Tuesday, 21st April.—Thermometer last night 84°, this morning 78°. While I went for the horses this morning, Gibson refilled the bags and kegs, and we made a very early start. I remarked to Gibson as we rode along, that this was the anniversary of Burke's and Wills's return to their depôt at Coöper's Creek, and then recited to him, as he did not appear to know anything whatever about it, the hardships they endured, their desperate struggles for existence and death there; and casually remarked, that Mr. Wills had a brother, who also lost his life in the field of discovery, as he went out with Sir John Franklin in 1845. Gibson then remarked, "Oh, I had a brother, who died with Franklin at the North-Pole, and my father had a great deal of trouble in getting his pay from Government." He seemed in a very jocular vein this morning (he was not often so; for he was generally rather sulky, sometimes for days together), and said to me, "How is it, in all these exploring expeditions, a lot of the people go and die?" I said, "Well, I don't know, Gibson, how it is; but there are many dangers in exploring, that may at any time cause the death of some of the people engaged in it besides accidents; but I believe want of judgment or knowledge or courage in individuals often brought about their deaths, but death was a thing that must occur to every one sooner or later." To which he replied, "Well, I shouldn't like to die in this part of the country anyhow;" in which sentiment I quite agreed with him, and the subject dropped. At eleven miles we were not only clear of the range, but had crossed to the western side of Lake Christopher, and were fairly enclosed in the sandhills; they were, of course, covered with triodia. Numerous fine casuarinas grew in the hollows between them; some small bloodwood-trees also ornamented the tops of some of the sandhills. At twenty-two miles on a west course we turned the horses out for an hour. It was very warm. There was no grass. The horses rested in the shade of a casuarina while we remained under another, and had some horse-flesh for dinner. I here discovered, that the bag with our supply of horseflesh in it held but a most limited quantity, to do two of us for a week, there being scarcely sufficient for one. Gibson had packed it up at starting from the Fort, and I had not previously seen it. The thermometer while we remained here rose to 96° in the shade. The afternoon was most oppressively hot, and I am confident the thermometer would have stood at 100°, if not more. Re-saddling, we got under weigh, and by night had got over an additional eighteen miles—making a day's stage of forty. The country was all sandhills. The Rawlinson-Range completely disappeared from view, even from the tops of the highest sandhills, at thirty-five miles. The travelling, though heavy enough, had not been so frightful, as I had anticipated, for the lines of sandhills mostly ran east and west, and by turning about a bit we got great numbers of hollows between them to travel in. Had I been going north or south, north-easterly or south-westerly, it would have been dreadfully severe. The triodia here reigns supreme, growing in enormous bunches and plots, and standing three and four feet high, while many of the long, dry tops are as high as a man. It gives the country the appearance of grassy downs; and as it is dotted here and there with casuarina and small patches of desert-shrubs, in general appearance the country is by no means displeasing to the eye, though frightful to the touch. No signs of the recent presence of natives were anywhere visible, nor had any of the triodia been burnt probably for many years. At night we got what we so pseudonymously call a "grass-flat," there being a few bunches of a thin and wiry kind of grass, of course white and dry as a chip. I never saw the horses eat more than one or two mouthfuls of it anywhere; at all events there was nothing else, and no water for them.

Wednesday, 22nd April.—Thermometer last night 84°, this morning 76°. The ants were very troublesome to me during the night, and I had to shift my quarters several times. Gibson did not seem at all affected by them. We were in our saddles immediately after daylight. I was in hopes that a few miles might bring about a change in the country; and so it did, but not a very advantageous one to us. At ten miles from camp the horizon became flatter, the sandhills fell off, and the undulations became covered with brown gravel, which was at first very fine. At fifty-five miles it became a little coarser, and at sixty miles it was evident the country was becoming firmer, if not stony. Here we turned the horses out, having come twenty miles. I found one of our large water-bags had leaked, and we had not as much water as I had anticipated. Gibson here preferred to keep the big cob to ride, instead of Badger, so after giving Badger and Darkie a few pints of water each, Gibson drove them back on the tracks about a mile, and let them go to find their own way and take their own time in returning to the Circus. They both looked extremely hollow and fatigued, and went away very slowly. We had poured some water into a flat hole at the Circus for them on their return, as they could not well drink from the reservoir. I had no doubt they would return all right, though sixty miles through such a country as this tells fearfully upon a horse, but now they could return slowly, and by night. We gave our two remaining riding horses all the water contained in the two large water-bags, except a quart or two for ourselves, which gave them a pretty fair drink, though not a tenth part of what they would have swallowed—they fed a little while we remained here. The day was quite too hot for traversing such a country; the thermometer, while we were here, standing at 96°, though we had not yet reached the hottest time of the day. We hung the two kegs with water (five gallons each) up in a tree with the pack-saddles, water-bags (now empty), and other gear. The supply in the kegs was, of course, intended to water the horses and ourselves on our return. Starting again, we made another twenty miles by night, the country still covered with small stones and thickly clothed with triodia; there were thin patches of low scrub at odd intervals. No view could be obtained to the west, all round us—north, south, east and west—was alike. The undulations forming the horizons were not generally more than seven or eight miles distant from one another; and when we reached the top of one, we obtained exactly the same view for the next seven or eight miles, and so on. The country still retained all the appearance of fine open grassy downs, and the triodia tops, waving in the warm breezes, had all the semblance of good grass. The afternoon had been very oppressive, and the horses were greatly disinclined to exert themselves, though my mare went very well. It was late by the time we encamped, and the horses were much in want of water, especially the big cob, who kept coming up to the camp all night, and tried to get at our water-bags, pannikins, &c. The extraordinary instinct of a horse, when in an agony of thirst, in getting hold of any utensil, that ever had water in, is surprising and most annoying. We had one small water-bag hung in a tree, and I did not think of it, when my mare came up straight to it, and took it in her teeth, forcing out the cork and sending the water up (which we were both dying to drink) in a beautiful jet, which, descending to the earth, was irrevocably lost. We had now only a pint or so left. Gibson was sorry he had exchanged Badger for the cob, as he found him exceedingly dull all the afternoon, which was not usual with him, as he was generally a most willing animal, though he only travelled at a jog, while my mare was a good walker. There had been a hot wind blowing from the north all day.

Thursday, 23rd April.—Thermometer last night 86°, this morning 74°. This was a most lamentable day in the history of this expedition, for it was the last I ever saw of Gibson. There was a most strange dampness in the atmosphere this morning, and it was cloudy to the west, the wind blowing from the south. The horses looked fearfully hollow this morning. We left the camp even before daylight, and as we had camped on the top of a rise, we knew we had seven or eight miles to go before another view could be obtained; we found the next undulation to be at least ten miles from camp, and upon reaching it I was glad to see at last some slight indications of a change. We were now ninety miles from the Circus-water, and 110 from Fort McKellar; the horizon in front of us, that is to say to the west, was still obstructed by another rise three or four miles away; but to the west-north-west I could see a line of low stony ridges; but they were ten miles off; to the south was also an isolated little hill six or seven miles off. I determined to go at once to the west-north-west ridges, when Gibson suggested the next rises to the west of us might reveal something better in front, so we went towards them, and reached them in five miles; there were others still in the way a little further on, and I went to them, it was three miles further; we had come ninety-eight miles from water. Here Gibson, who always rode behind me, called out and said that his horse was going to die or knock up (which are synonymous terms in this region). We had here reached a point, where at last a different view was presented to us, and I believed a change of country was at hand; for the whole western and down to south-western horizon was broken by lines of ranges, being most elevated in the south-western direction; they were all notched and irregular, and no doubt formed the eastern extreme of a more elevated and probably mountainous region to the west of them. The ground we now stood upon, and for a mile or two past, was almost a stony hill itself; and now for the first time in all the distance we had travelled, we had reached a spot where water could run during rains, though of course we had not seen any place yet where it could lodge. Between us and the hilly horizon to the west the country seemed to fall into a long valley, it looked dark and seemed to have timber in it; and here for the first time since leaving the Rawlinson-Range we saw that the ground that we were on, and that for some miles around, had been burnt by the natives, though not recently. The hills bounding the western horizon were between twenty and thirty miles away, and it was with extreme regret that I was compelled to relinquish a further attempt to reach them. Oh, how ardently I longed for a camel; how ardently I gazed upon this scene! At this moment I would even my jewel eternal have sold for power to span the gulf that lay between; but it could not be; situated as I was I was compelled to retreat, and the sooner the better. On desire of Baron von Mueller, especially also in commemoration of their recent marriage, I named this farthest-sighted feature the "Alfred and Marie Range," in honour of their Royal Highnesses the Duke and Duchess of Edinburgh. There is scarcely any doubt, that these ranges shut out from the view other and higher ones to the west. Gibson's horse having shown signs of giving in, had placed us both in a great dilemma; indeed it was a most critical position to be in. We therefore returned upon our tracks at once, and had scarcely gone a couple of miles before the cob refused to carry his rider any further; we drove him on foot another mile, and he lay down to die. My mare, the Fair Maid of Perth, was only too willing to return. She now had to carry Gibson's saddle and things, and we pushed on as fast as we could, walking and riding by turns. The cob, of course, died where he fell; not a second thought could be bestowed upon him. When we got back to ninety miles from the Circus, and thirty miles from the kegs, I was walking, and having concluded in my own mind what course to pursue, called to Gibson to halt, till I came up to him. We were both excessively thirsty, for walking had made us so, and we had scarcely a thimbleful of water. However, what we had, we took a mouthful of each, which finished the supply, and then I said—for I couldn't speak before, "Look here, Gibson, you know we are in a most terrible fix, and only one horse; therefore only one person can ride, and one must remain behind. I shall remain; and now you listen to me. If the mare does not soon get water, she'll die; therefore go right on, get to the kegs if possible to-night, and give her water, and now since the cob is dead all the more can be given her. Give her an hour or two's rest, and then try and get over a few miles more by morning, so that early to-morrow you will again sight the Rawlinson-Range. Stick to the tracks and don't leave them, but leave as much water as you can afford (after supplying yourself and filling your bags and watering the mare) for me, and remember that I depend upon you to bring me relief. Rouse Mr. Tietkens, get fresh horses and more water-bags, and return as soon as you possibly can. I shall endeavour to get down the track also." He then said, if he had a compass, he thought he could go better at night. I knew he didn't understand much about compasses, and I was loth to part with mine, as it was the only one I had. However, I gave it to him, and he departed. I sent one final shout after him to stick to the tracks, to which he replied "All right," and the mare carried him out of sight immediately. I walked slowly on, and the further I walked the more thirsty I became. I had thirty miles to go to the kegs, and I should not reach them until late to-morrow at the rate I was travelling, and I did not feel at all sure, I could keep on at that. The afternoon was very hot; but the evening became more cool. I continued upon the tracks until the moon went down, and then had to encamp reasonably cool, though choking for water. Having no other clothes but shirt and trousers, I was compelled before morning to light a fire. How I longed again for morning. I hoped Gibson had reached the kegs, and that he and the mare were all right. I could not sleep for thirst; I could only lie and wait for morning, wishing this planet would for once accelerate its movements, and turn upon its axis in twelve instead of twenty-four hours, or rather that it would complete its revolution in six hours.

Friday, 24th April (till 1st May).—Thermometer last night 78°, this morning 68°. As soon as it was light this morning, I was again upon the horse-tracks, and reached the kegs about the middle of the day. Gibson had been here, watered the mare and gone on; he had left me about two and a-half gallons of water in one keg; and it may be imagined, how glad I was, to get a drink; I could have drank the whole quantity in an hour, but I was compelled to economy, for I could not tell how many days would elapse, before assistance could come; it could not well be less than five days—it might be more. After quenching my thirst a little, I felt ravenously hungry, and on searching among the bags all the food I could find was eleven sticks of smoked horse-flesh, averaging about one and a-half ounces each. I was rather staggered to find, that I had little more than a pound weight of meat, to last me till assistance came. However I was compelled to eat some at once, and proceeded to devour two sticks raw, as I had no water to boil them in. After this I sat in the shade, to reflect on the precariousness of my position. I was sixty miles from water, and eighty from food; my messenger could not well return before six days, and I began to think it highly probable, I should be dead of hunger if not thirst before anybody could possibly arrive. I looked at the keg, it was an awkward thing to carry empty; there was nothing else to carry water in, as Gibson had taken all the bags. The keg when empty with its straps and rings weighed 15 lbs., and now it had 20 lbs. of water in it, I could not carry it without a blanket for a pad for my shoulder (I had left my blanket here when going out), so that with my revolver and cartridge pouch, and other things on my belt, I staggered under a weight of about 50 lbs., when I put the keg on my back; I had fourteen matches only, Gibson had all the others. After I had thoroughly digested all points of my situation, I concluded, that if I did not help myself, Providence wouldn't help me either. So I started away with the keg on my back, bent double with it, and could only travel so slowly, that I thought it scarcely worth while travelling at all; I got so thirsty at every step I took, I longed to drink every drop of water I had in the keg, but I restrained myself. The thermometer did not rise quite so high to-day as yesterday, it going up only to 90° in the shade. I only got three miles away from the kegs, and to do that I travelled most of it in the moonlight, after the sun went down. The next few days I shall pass over with the remark, that as long as water lasted in the keg I carried it, and averaged about five miles a day. To people, who cannot comprehend this country, it may seem absurd, that a man could not travel faster than that. All I can say is, there may be men who would do so; but most men in the position I was in, would simply have died of both hunger and thirst, for by the second day my horse meat was all gone. I had to remain in what shade I could find during the day, and I travelled by moonlight at night. At fifteen miles from the kegs the two loose horses, we had turned away from there, had left the main line of tracks, which of course ran east and west; these two horses left them at a slight incline, going on about E.S.E. Gibson, I was grieved to see, had gone in the loose horses' tracks; but I felt sure he would and they also would return to the main-line, and I therefore could not investigate any further in my present position; so after following them about a mile I turned again to the main line of tracks, anxiously looking at every step to see, if Gibson's horse-tracks returned into them. They never did, nor did the loose horse-tracks either. Generally speaking, whenever I saw a shady tree there was an enormous bulldog ants' nest under it, and I was prevented from sitting in its shade. On the 26th I almost gave up the thought of walking any further; for the exertion in this dreadful country, where the triodia was almost as high as myself, and as thick as it could grow, was quite overpowering, and being now almost starved I felt quite light-headed, and after sitting down on several occasions, when I got up again my head would swim round, so that I fell down many times, and was often oblivious for more than a quarter of an hour; and being in a chronic state of thirst the whole time, my general plight was dreadful in the extreme. After this my great hope and consolation was, that I might now soon meet the relief party, and might see them at any turn of the track; but where was the relief party? I often asked myself, and echo could only answer—where? On the 29th I had emptied my keg, and was still over twenty miles from the Circus; but in this *April's ivory moonlight* I plodded on desolate, yet all undaunted, and reached the Circus-water just at dawn of day. Oh, how I drank; oh, how hungry I was; and I thanked Providence that I had so far at least relieved myself from that howling wilderness, for I was now once more upon the Rawlinson-Range, though still twenty miles from home. There was no sign of any tracks of anyone having been here since I left it; the water all but gone. I wondered, what had became of Gibson; he had certainly never come here, and how could he reach the Fort without coming here—that was the question; but I was in such a miserable state of mind and body, that I ceased to add any more vexatious speculations as to what had delayed him. I remained here continually drinking and drinking, until 10 a.m. The ground was now of course all stones, my feet were very sore, and I could only go at a snail's pace over them. So weak had I become, that by late at night I had only accomplished eleven miles, and lay down about five miles from the Gorge of Tarns, again choking for water. Most of these days the thermometer had stood at 92° in the shade. I omitted to mention, that just as I left the Circus I picked up a small dying wallaby, whose mother had thrown it from her pouch. It weighed about two ounces, and was scarcely furnished yet with fur. The instant I saw it, like an eagle I pounced upon it and ate it raw, dying as it was, fur, skin and all. The delicious taste of that creature I shall never forget. I only wished I had its mother and its father, to serve in the same way. On the 1st of May, at 1 o'clock in the morning, I was walking again, and reached the Gorge of Tarns long before daylight, and could now indulge in as much water as I desired, but it was exhaustion I suffered from. My reader may imagine with what intense feelings of relief I stepped into Fort McKellar at daylight and awoke Mr. Tietkens, who stared at me, as though I had been one, newly risen from the dead. I asked him, if he had seen Gibson, and to give me some food. I was appalled to hear, that Gibson had never reached the camp, nor had any of the loose horses come back; and it appeared, that I was the only one of six living creatures—two men and four horses—that had returned, or were ever likely to return, from that desert, it being now nine days since I last saw Gibson. Mr. Tietkens told me that he had been in a great state of anxiety during my absence, and had only returned himself a few hours ago from the Circus, where he had planted some food and marked a tree; that he had seen my foot-tracks, but could not conclude they were so recent, as he thought they were left, when Gibson and I were here together. It seems we had missed one another somewhere on the range. We were both equally horrified at Gibson's mischance. I told him just what had occurred out there, and how Gibson and I had parted company, and we could only conclude, that he must be dead, or he would long before have returned. The mare certainly would have carried him to the Circus, and then he must have reached the depôt; but I now of course saw, that he had gone wrong, had lost himself, and must now be dead. I was too exhausted and prostrate to think of moving from the camp, to search for him to-day, but I determined to start to-morrow. Mr. Tietkens got everything ready, while I remained in a state of semi-stupor the whole day. I was cramped with pains in all my joints, pains in the stomach, and violent headaches, these being probably the natural result of having a long-empty stomach suddenly filled. Jimmy was much pleased to see me, and I suppose grieved at the loss of Gibson, though I do not think they were much attached to one another. Gibson's loss and my struggles of course now formed our only topic of conversation for the whole day, and it naturally shed a gloom upon our spirits. Here we were isolated from civilisation, and hundreds of miles away from our fellow-creatures, and one of our small party had gone from us. It being now nine days since I parted with him, it was not very likely that he was still in existence in that fearful desert, as no man would or could *stay* there *alive;* he must be dead, or he would have returned, as I did, only much sooner, as the mare he had would carry him as far in a day, as I could walk in a week in this country. Thermometer to-day 90° in the shade. This afternoon was very cloudy and threatened rain. I was now however in hopes that none would fall, as it would only obliterate the tracks of our lost companion in the sandy desert, in which he had strayed.

Saturday, 2nd May.—Thermometer last night 72°, this morning 64°. Before daylight this morning we were awoke by the sprinkling of a light shower of rain, the rain was too light to be of the slightest use in leaving any water, but yet it continued so long, that I felt sure we should have some trouble in following Gibson's tracks after it. The rain ceased about seven o'clock, leaving everything damp and dirty, without having done the slightest good, for it left no water even on the bare rocks. Mr. Tietkens and Jimmy got all the things we required and the horses—I was so weak and unfit to go away, that I could do nothing. We took three pack-horses to carry water and two riding horses; but starting very late, we did not reach the Circus by night. The day was much cooler after the slight rain, thermometer not rising higher than 80°; we camped on a dry creek at thirteen miles. Mr. Tietkens had not been attacked by the natives, while I was away, but they were always in close proximity, and their presence could always be detected by their perpetual fires, for in these regions no game can be got —it appears—without the aid of fire. I was very loth to leave Jimmy Andrews alone in the camp, and under any ordinary circumstances I would not have done it, but was now compelled to, and though I dreaded and expected his being attacked, still I hoped it might not occur.

Sunday, 3rd May.—Thermometer last night 70°, this morning 62°. The horses were actually in sight of the camp this morning. I was dreadfully stiff, and yesterday's short ride had fatigued me greatly. We had carried all our water-bags, three pairs full, from the depôt, as the water at the Circus was now all but gone—reaching it early we turned the horses out for a few hours, while we topped up the bags; they held a good quantity of water at starting, averaging sixteen gallons each pair, but a quantity got squeezed and jolted out. It was late in the afternoon, when we left the Circus, which was now dry—with the exception of a few quarts; there was good grass on the edges of Lake Christopher, and we camped there for the benefit of the horses, as they were not likely to get any more; we had come nine miles from the Circus. The thermometer rose only to 80° in the shade, and the evening set in cool, indeed quite cold, which I was exceedingly glad of on account of the horses.

Monday, 4th May.—Thermometer last night 58°, this morning 48°. There was a great change in the temperature this morning. I was exceedingly ill this morning, and had to be helped on to my horse at starting. We were now on the main line of tracks. I was obliged to go back as far as the kegs, as I had left several things, such as one keg, saddle, bridles and other things indispensable to us. They were sixty miles from the Circus. Gibson's tracks turned off the main line fifteen miles nearer this way; but, having to go to the kegs, I intended to follow his trail as we returned. Yesterday's shower had deadened the tracks considerably. We only got thirty miles to-day, as I could not ride any further. The weather was fine and cool, the thermometer not rising higher than 80° in the shade. The tracks now were not very easy to follow; but, by steering west, as I had formerly done, we kept so close to them that we saw them here and there; and, as Gibson and I, in going out, had burnt the triodia occasionally, these places served as landmarks; and then, again, I practically remembered a good many spots. We camped on what we call a grass-flat for the night, giving each horse three-quarters of a bucketful of water; and, there being five horses, it emptied one pair of bags.

Tuesday, 5th May.—Thermometer last night 58°, this morning 48°. Another splendid cold night. I was very cold in the night: my bedding I had left at the far keg. We reached the sixty-mile keg at two p.m. to-day, and gave the horses an hour's rest, while Mr. Tietkens went on a few miles further, to a tree where Gibson had left his saddle, to lighten the mare on the day we parted. We gave Mr. Tietkens's horse Diaway a bucketful of water, and he went for the saddle, while I remained and gave each of the others three-quarters of a bucket. Mr. Tietkens returned before sundown with the saddle. Diaway had now to be rested for an hour; but none of the horses would attempt to eat any of the grass, that was there. When the moon rose we departed; but I was in too great pain to continue long in the saddle; and we only got nine miles back on the tracks. The day had been delightful, the thermometer not rising higher than 76° in the shade.

Wednesday, 6th May.—Thermometer last night 58°, this morning 42°; fine cold night, and excellent weather for the horses; and, now I have recovered my bedding, I enjoyed it all the more. At seven miles after starting we reached the fork of the two lines of tracks—that is to say, the spot where the two loose horses had left the main line, and upon which Gibson had gone also; I hoped to come upon some traces of him before night. We had already been three nights without water, and our supply in the bags was now all but gone; we fortunately had fine cool weather. As long as Gibson's track remained upon the others the tracking, though not very easy, was practicable enough; but that unfortunate and foolish man had left these tracks, and gone away in a far more southerly direction, having the most difficult sandhills to traverse now at right angles. He had burnt a patch of triodia, where he left the other horse tracks, and must have been under the delusion, that they were running north, and that the main-line of tracks must be on his right hand, or south of him; and whether he made any mistake in steering by the compass or not is impossible to say; but instead of going east, as he should, he actually went south or very near it. In consequence of the small reptiles, such as lizards, which always scratch over all horse-tracks during night in this country, and the slight rain we had the other morning, combined with the shifting nature of the sand, we could make but poor headway in following the track, and it was only by one of us walking, while the other brought on the horses, that we could keep it at all; and our pace was so slow travelling in this manner that, although we did not halt at all during the day, by night we had not been able to trace him more than twelve or thirteen miles; up to this point there seemed no apparent diminution of the powers of the animal he bestrode. We camped again for the fourth night, without water, upon the tracks, not being able to follow it in the moonlight, and we gave our horses all our remaining stock of water. We began to think, that our chance of finding the remains of our lost comrade were very slight. The thermometer rose only to 80° in the shade. Our reflections upon Gibson's loss were gloomy indeed; and I was sorry to think, that the unfortunate man's last sensible moments must have been embittered by the thought, that having lost himself in the capacity of messenger for my relief, I too would become a victim to his mishap. I called this terrible region, that lies between the Rawlinson-Range and the next permanent water, which may eventually be found to the west, "Gibson's Desert," after this first white victim to its horrors. Gibson having had my horse, rode away in my saddle, and my field-glasses were attached to it; but everything was gone, man and horse were alike swallowed in this remorseless desert.

Thursday, 7th May.—Thermometer last night 56°, this morning 42°. I was most thankful for the continuance of this fine cool weather, or we could not have remained away from the water so long. Mr. Tietkens and I consulted together, and could only agree, that unless we came across Gibson's remains by mid-day, we must of necessity retreat, otherwise it would be at the loss of fresh lives—human and equine; for, as he was mounted on so excellent an animal as the Fair Maid of Perth (on account of whose excellence I had chosen her to ride), it seemed quite evident, this noble creature had carried him only too well, and had been ridden literally to death, having carried the rider too far from water ever to return, if he had known where it lay. What actual distance she had gone from here, of course was impossible to say. I felt more at ease in walking along the tracks than riding; but we went so slowly, mile after mile, rising sandhill after sandhill, until twelve o'clock, not having followed them then more than seven or eight miles since morning, we could not reach the Circus by night; for we were nearly fifty miles from it, and I was in great anxiety as to whether any water would be there at all on our return; therefore we had to return, having been quite unsuccessful in our melancholy search. I may here remark that, when left, Gibson's tracks were going in the direction of, though not straight to, the dry ridges that Jimmy Andrews and I had visited in February, which were in sight, and no doubt Gibson imagined, they were the Rawlinson-Range, and he probably ended his life amongst them; but we could go no farther now. We made a straight line back for the western end of Rawlinson-Range; and, as we continued on until nearly morning, we did not stop until the western edge of Lake Christopher was reached. Being the fifth night from water, the horses were very bad. We were now only ten miles from the Circus, and hoped to reach it early again in the morning. The day had been fine and cool, thermometer 80° in the shade. I should have remarked, that before reaching the dry lake, and several miles south-westerly from it, by the merest chance, that is to say—by the horses we were driving running along them, we crossed and saw the tracks of the two loose horses, we had turned away from the kegs; they were making too southerly ever to reach the Rawlinson-Range. Where these unfortunate brutes wandered to and died can never be known, as it would cost the lives of men simply to ascertain.

Friday, 8th May.—Thermometer last night 64°, this morning 56°; a warmer night than the few preceding ones. On reaching the Circus we found, that though there was still a small quantity of slime and liquid dirt, it was actually useless for our horses, which were now in a very bad way, and consequently travelled exceedingly slow. We put into the Gorge of Tarns, which we did not reach till late, and then of course our horses were satisfied, though this is a rather awkward place for horses to water. Reaching the depôt late at night, I was almost more exhausted than when I had walked into it last. Thermometer to-day rose only to 80° in the shade. We found Jimmy Andrews all right in the camp, and heartily glad he was of our return; for it must be remembered, that Jimmy is but young, and to be left alone in so lonely a spot, with the constant dread of hostile attacks from the natives, would be by no means pleasant to anyone. When we arrived we found, that all our stock of poor old Terrible Billy was gone, and it was necessary to kill another horse. Mr. Tietkens had partially erected another smoke-house, and to-morrow we must work at it again; the affairs of the dead must give place to those of the living, though I could not endure the thought of leaving Gibson's last resting-place unknown.

Saturday, 9th May.—Thermometer last night 70°, this morning 66°; a warm night and cloudy morning. I could not rest in my mind without another attempt to discover Gibson; but now the Circus waters were gone, and it would be useless to go from here with any hopes of following the tracks, for where we had left them was nearly seventy miles from here, and by the time we got there again it would be time to return. In the early part of the day we got sticks and logs, and erected a portion of the smoke-house, and Jimmy got the horses. I then determined to go with Mr. Tietkens to the Gorge, where he and Gibson had found the rock waterhole, and which they said was unapproachable. I was determined to see, if it were available, so we delayed the killing of another horse until our return, and in consequence we had to consume nearly all our remaining stock of flour. In the afternoon we took five more horses—two to ride, and three to carry water—but it being late, we had to camp ten miles on this side of the range. Thermometer only 78°.

Sunday, 10th May.—Thermometer last night 64°, this morning 54°. A fine cool night. We had some difficulty in getting over the range. Having gone up a creek-channel five miles from where we encamped, arriving at the other side, we travelled up it for eight miles to the gorge, Mr. Tietkens had found. There was certainly some water there, but it really was, as Mr. Tietkens had said, utterly unapproachable, and too far to carry water to horses; so I was compelled to relinquish any further search after Gibson, for the distance now from available water to where we left the tracks was so great, that no more efforts could be made. We searched among two or three other gorges, but they were all dry, so we gave up the attempt, and returned towards home. I first called the most western hill in the Rawlinson-Range "Mount Forrest," after Mr. John Forrest, the well-known explorer of Western Australia, who is now in the field; and the most western watercourse that issues from this range, "Forrest's Creek," after his brother, Mr. Alexander Forrest, who has also been the leader of exploring expeditions in this Province. We only got over to the north side of the range again by night, and encamped. Thermometer 80° in the shade to-day.

Monday, 11th May.—Thermometer last night 66°, this morning 54°. Another fine cool night. Getting our horses very early this morning, and not having far to go, we arrived again at the depôt before twelve o'clock. Jimmy this time truly rejoiced at our return, for it appears, that actually while we were getting ready to depart on Saturday, a whole army of natives had been hidden by the rocks above the camp, waiting and watching until we departed; and as soon as we were well out of sight and sound, had commenced an attack upon poor Jim; and from his statements, it was only by the continued use of Snider-bullets —of which fortunately I had a good supply—that he could keep them out of the camp at all. They threw several spears and great quantities of stones down at him; and it was only by his dodging and hiding behind the pallisades, that he escaped death. He appears however to have been far too flurried to have hit many of them; as I don't think either he or Gibson ever saw a gun before they came with me. Every time he fired, they would simply bob down their heads, and whistle in imitation of the sound of the passing bullet, which, as they couldn't see, and didn't feel, they held in great contempt; all the afternoon the contest lasted, and they tried to set fire to the things in the camp. At night it seems one of them got wounded, and most of them departed, but Jimmy, with the little dog, watched all night, and before daylight on Sunday they all came again and recommenced their hostilities. Jimmy was now becoming desperate; one individual came, as he thought, to give him his quietus, and commenced dancing about the camp and pulling over many of the things, when Jimmy suddenly caught up a shot-gun with cartridges in it (of which I kept about a dozen for defence), and before the fellow could get away, he received the full charge about his body. Jimmy said he bounded in the air, held up his arms, and shrieked, screeched, screamed and finally ran off with all the others, some before and some after him; and from that time they had not troubled him any more. I gave the lad great praise for his spirited and plucky action, for it was a predicament, that no one could desire to be placed in; in fact, it was a most fortunate escape from most probably a cruel death, if indeed these animals would not have actually eaten him.

The day was delightfully cool, the thermometer only rising to 78° in the shade. We finished the smoke-house this afternoon, and got the victim we were going to slay, and tied him up all night. The victim this time was Tommy. I had brought this horse with me from Victoria, and had him out upon my last expedition; on this expedition I had been compelled to repurchase him at a startling advance upon his original Victorian price. He was now very old and very poor, two coincidences, that can only be thoroughly comprehended by the antiquated of the human race; and for my part I would rather be eaten by savages in my youth or at mature age, than experience such calamities at an advanced period. Tommy did not promise much oil.

Tuesday, 12th May.—Thermometer last night 66°, this morning 56°. I shot poor Tommy early, and we got him into the smoke-house (with the exception of what we kept fresh) by afternoon. We had to boil every bone in his body, to get sufficient oil to fry steaks with, and the only way to get one's teeth through them, was to pound them well before cooking. I wish I had a sausage-machine. The day was again cool, the thermometer rising to 78° in shade.

Wednesday, 13th May.—Thermometer last night 70°, this morning 68°. I previously mentioned, that I had a plan for penetrating this western desert, when the weather changed. The plan was this:—I would have cut up and made available every particle of material in my possession, to make waterbags of, and before Gibson's loss I could have carried nearly 100 gallons of water. I had intended to have packed water out to various distances until I had, by being kept supplied by Mr. Tietkens and Gibson from behind, either gone 160 or 170 miles or found water, and then by packing all the worst horses with water, by sacrificing their lives as the water was drunk, to get the others over even that distance; and I don't think the next water is any further. Even with this plan only one man could ever remain in the camp, and now that Gibson was dead it could not be attempted, as two were actually required to keep moving and keep up the supplies at the various stages, while I would be always in advance after the first stage or two; so after admitting to myself, that I was actually unable to accomplish more with the means at my disposal, I most reluctantly (and it will be believed sorrowfully) decided to return. Our stores and clothes were now gone, and we had nothing but horseflesh; still if Gibson hadn't died, I should not have abandoned this region so soon. Though of course neither Mr. Tietkens nor Jimmy could receive my intimation otherwise than with pleasure, still they were both ready and both willing, and as anxious, I may say, as I was, that our undertaking should be successful; but as we stood, in our present circumstances, nothing more could be done. We set to work to shoe some of the horses. As soon as Tommy is smoked we shall depart. Tommy proved to have more flesh upon his bones than I anticipated, and he may do us for a month.

Thursday, 14th May.—Thermometer last night 74°, this morning 68°. Still at the Fort smoking Tommy, and shoeing horses. Thermometer 86° in shade.

Friday, 15th May.—Thermometer last night 74°, this morning 68°. Smoking and shoeing. Thermometer 90°; the day very warm.

Saturday, 16th May.—Thermometer last night 78°, this morning 70°; a warm night. Thermometer rose to 90°; atmosphere very sultry.

Sunday, 17th May.—Thermometer last night 80°; this morning 70°. Another warm night; it being sultry and cloudy in the morning. We also had a few drops of rain. Oh, how I wish for rain, then we could push out west, as the Alfred and Marie Range, I had sighted to the west, would certainly have water after rains; but none fell. Every drop in this singular region seems meted and counted out, yet there are the marks of heavy floods on every watercourse; the question of when did the floods occur which caused these marks, and when, oh, when! will such phenomena occur again, is always recurring to me. The climate of this region seems most extraordinary, for both last night and the night before we could all lie on our blankets without requiring a rag to cover us; such a thing never occurred to me in May in any other part I have visited, and I cannot determine, if it is a peculiarity of this region, or if it is an unusual season throughout this half of the continent; for with the exception of a few thunder showers, which fell in January, not a drop of rain to leave water has fallen since I left the Telegraph-Line; and there the country was in almost a state of drought, though we had a shower or two just at starting. The thermometer to-day rose only to 76°, and the day was cool and cloudy. I cannot leave this singular spot without a few remarks on its peculiarities and appearance—for its waters are undoubtedly permanent, and may be useful to future travellers; in the first place it bears 12° E. of S. from the highest ridge of Mount Destruction, in the Carnarvon-Range; that mountain, however, is partially hidden by the intervening low hills, where Mr. Tietkens' riding horse, Bluey, died. I always called it Bluey's Range; and this depôt is amongst a heavy clump of fine eucalypts, which, however, are only thick for about a quarter-of-a-mile. From underneath this clump a fine strong spring of the purest water for ever flows, and just opposite to our camp it opens out into a fine little basin with a stony bottom; the surface-water only exists for half-a-mile. This is the most western permanent water discovered by me, that is to say the most available, for the Gorge of Tarns, four miles further west, is no doubt permanent, but it is not nearly so convenient as this. The grazing capabilities of the country about here are very poor, and my horses have only just existed, since I left the Pass.

Monday, 18th May.—Thermometer last night 68°, this morning 58°. A fine cool night, and there seems some likelihood now of a permanent change of the weather for the season; rains we have long ceased to expect. Nature, in her own good time will, I suppose, reanimate this sinking region with her showers.

Tuesday, 19th May.—Thermometer last night 66°, this morning 58°; day cloudy. Getting everything ready for our departure; thermometer 78° in the shade.

Wednesday, 20th May.—Thermometer last night 60°, this morning 50°. This day month Gibson and I started away for the west; thermometer 70°. Three natives came up near the camp to-day, but as they or their tribe had so lately attacked the camp, I had no very friendly feelings towards them, though we had a peaceable interview, and the only information I could glean from them was that their word for travelling, or going or coming, was *Peterman.* They pointed to Mount Destruction, and intimated that they were aware we had *Peterman'd* there; that we had *Peterman'd* both from the east and to the west; everything with them was *Peterman.* It is singular how identical the word is in sound with the name of the celebrated Dr. Petermann, the geographer.

Thursday, 21st May.—Thermometer last night 62°, this morning 52°; cold windy night. In the afternoon we commenced our retreat, and finally left Fort McKellar, where my hopes had been as high as my defeat signal. We only got fourteen miles on our road to the Pass.

Friday, 22nd May.—Thermometer last night 58°, this morning 48°; cold and windy night. We reached the Pass once more early, and camped up close to the freshwater-springs. Mr. Tietkens had planted a small patch of splendid soil (and poor Gibson had done the same at Fort McKellar), with all kinds of seeds, but the only thing that came up well here was the maize, and that looked splendid, nearly three feet high. Thermometer 68°.

Saturday, 23rd May.—Thermometer last night 58°, this morning 46°; a cold fine night and delightful day, thermometer 68°. We rested at this most delectable spot for another day. I greatly liked this place; there was a long line of fine eucalyptus-timber, and an extensive piece of ground all covered with green rushes, which was very pretty. There never had been any ants here, and altogether it was a most desirable explorer's camp. I do not suppose I shall ever forget Sladen-Water and the Pass of the Abencerrages. It was an excellent place also for the horses, and they all got fat here.

Sunday, 24th May.—Thermometer last night 62°, this morning 48°. We gave a last farewell to Sladen-Water, and departed this morning for the east; and I now travelled to the eastward on the line, explored in March, that is to say, passing and obtaining water at all the following places:—Gill's Pinnacle, MacBain's Springs, Harry Hull's Springs, Louisa's Creek, the Chirnside, the country, as I said before, being very excellent. The crescent-shaped and wall-like range, running from Weld's Pass to Gill's Pinnacle and beyond, I named on request of my friend Baron von Mueller the "Schwerin Mural Crescent," and a magnificent pass through it, "Vladimir Pass," in honour of Prince Vladimir, son of the Emperor of Russia, lately married to the Princess of Schwerin.

Monday, 1st June.—Thermometer last night 54°, this morning 48°. We made a detour from our former line, by going a little more northerly, to avoid some rough hills which we had formerly crossed, and found another splendid little watering place, where several creeks joined, and then went down through a rocky glen to the north. There was plenty of rock- and sand-water here, and it was a very pretty place. I called this most excellent little spot "Winter's Glen," and the main creek of the three in which it lies "Irving-Creek," after Wm. Irving Winter, Esq., of Stanhope, a subscriber to the fund for my expedition. This water may easily be found, from its bearing from a high, long-pointed, and abruptly-ending (to the west) hill, which I named "Mount Phillips," after my very kind and oldest friend in South Australia, Dr. Phillips, of North-terrace. This is a very conspicuous mount in this region, being, like many of the others named on this line, detached, and yet forming a part of the long northern wall, of which the Petermann-Range is formed. This mount can be seen distinctly from Mount Olga, though it would be seventy miles away (and from whence it bears 4° north of west). The water at Winter's Glen bears west from Mount Phillips, and is distant four miles from it. We were now again on the territories of South Australia, having bid farewell to her sister State, and had turned our backs upon that peculiar province of the sun, the last of Austral lands he shines upon.

Tuesday, 2nd June.—Thermometer last night 46°, this morning 34°. A very cold, clear, calm night, with a slight hoar-frost in the morning. I now wished to make for Glen Robertson, at Casterton-Creek, which I visited and named on the 15th March. We had to make a more circuitous line than formerly, but found better country to traverse, and reached the glen early. There was yet another detached hill in the northern line, and which is the most eastern of the Petermann-Range; it bore only two or three degrees east of north from this water. I named it "Mount McCulloch," after the Hon. A. McCulloch, of Gottlieb's Wells, at whose stations, and that of his son's, I have been most kindly and hospitably entertained. This mount is also visible from Mount Olga, and bears 2° north of west from it.

Wednesday, 3rd June.—Thermometer last night 56°, this morning 38°. We remained here all day. Several natives made their appearance, and made signal-fires for others. One wretch sneaked so close upon us down the rocks behind the camp, that he almost touched us, before he was discovered. He got away before we could catch him. No doubt several others were near him, and that they intended to spear us, for this is the place (or close to it) where we were so beset by them in March.

Thursday, 4th June.—Thermometer last night 56°, this morning 50°. Mount Olga from here bore nearly east, and Stevenson's Peak bore 33° south of east. We had got some rock-water at that place going out, and as it was much nearer than Mount Olga, I determined to visit it again, hoping a chance shower might have left a drop of water there lately. In nine miles after starting we came to and crossed another large, wide creek, running up to the north. We were some distance from its head, and there was no water, where we crossed it. I called this the "Armstrong," after W. Armstrong, Esq., of Hexham, Victoria, a subscriber to the fund for my expedition. At twenty miles I found another fine little glen, with a large rockhole, and water in the sand. I called this "Wyselaskie's Glen," and the creek, in which it exists, the "Hopkins," after J. D. Wyselaskie, Esq., of the Hopkins, another subscriber to the fund for my expedition. Here we encamped. It was a very pretty spot, and the grass good.

Friday, 5th June.—Thermometer last night 60°, this morning 60°; a warm, cloudy night. We returned to Mr. Gosse's and my old camp, and found enough water for only three horses, but at another spot found sufficient for all the rest. The other spot is half-a-mile west of the gorge, and over a hill; it is a place which holds water a long time. We pushed on now for Mount Olga, and camped in casuarina- and triodia-sandhills without water.

Saturday, 6th June.—Thermometer last night 54°, this morning 36°. Cold and very windy night, though it became calm after the sun arose. My only remaining thermometer is not graduated below 36°, and the mercury descends into the bulb at that figure. The day was cold, and windy in the afternoon. It was very late at night when we reached Mount Olga, as two horses straying had delayed us. Of course, I could not be sure of getting water, but was well pleased to find the little purling brook gurgling along its rocky bed; and all the little basins were full—the water, as when I last saw it, ending where the solid rock fell off. The country all round was excessively dry, and the grass all withered, though in the channel of the creek there was some a little green.

Sunday, 7th June.—Thermometer last night 54°, this morning 42°. I had a great desire from here to penetrate straight east to the Finke, as a considerable distance upon that line was yet quite unknown, and I determined to attempt it to-day. However, we remained here for the benefit of the horses; one of them, Formby, was unwell and rather troublesome. We were now again almost at the end of our stock of Tommy, and Formby must be the next victim; but we can yet wait a week, so I delayed the execution.

Monday, 8th June.—Thermometer last night 56°, this morning 50°. Another windy night, and this morning it threatened rain; but as usual none fell. Thermometer 62° in the shade. We remained here again to-day, though I went round to the northward, to inspect that side of this singular and utterly unclimbable mountain. Our camp was at the south face under the highest mound of the mountain. On the west side I found another running spring, with some much larger rock-basins than at our camp; of course the water ceased running, where the rock ended. Round to the north side I found a much stronger spring, and in a much larger channel. I went completely round the mountain and returned to the camp. The extraordinary appearance of this feature will never be out of my remembrance; it is, I should say, decidedly of volcanic origin, belched out of the bowels and on to the surface of the earth by the upheavings of subterraneous and subaqueous fires, and cooled and solidified into monstrous masses by the gelid currents of the deepmost waves of the most ancient of former oceans. As I said before, it is formed of mixed and rounded stones into rounded shapes; but some upon the eastern side are turreted, and are almost pillars, except that their thickness is rather out of proportion to their height. The highest point of this mountain I made 1485 feet from the surrounding country. To the eastward from here, and no great distance off, is another most singular looking mount (and which I also saw in my former expedition, but from the distance it was looking too small to engage much attention, Mount Olga to the west of it was so much higher), and similar I suppose to this. Beyond that again, and still further to the east, and a very long way off, was another range; but very indistinct from distance. Mr. Gosse's dray-track had evidently come from the nearest hill, and to-morrow I intended to visit it also.

Tuesday, 9th June.—Thermometer last night 56°, this morning 46°; windy night; wind from east. This morning we went off to the near bare-looking mountain to the east. It is twenty miles from Mount Olga. We saw Mr. Gosse's dray-track several times, and on reaching it saw his camp also. There was no water where Mr. Gosse had camped, but round another turn of the hill we found a very fine deep pool lying in the sand, and under the rocky face of the mount. This was upon the south side. There was also a fine, deep and shady cave here, ornamented in the usual aboriginal fashion, or indeed in an unusual aboriginal fashion, for there were two marks upon the walls three or four feet long of parallel lines, with spots between them. I considered that these marks were the native representations of their notions of depicting Mr.

Gosse's dray-track, with the horses' tracks between, though they may have been there when Mr. Gosse visited this mount. Since I returned I find, that Mr. Gosse named this mount, Ayers-Rock, and of course by that name it will for ever be known. The appearance and outline of this rock is most imposing, for it is simply a mammoth-monolith, that rises out of the sandy soil around, and stands with a perpendicular and totally inaccessible face at all points, except one slope near the north-west end, and that is at best a most precarious climbing ground, with a height of more than a thousand feet; down its furrowed and corrugated sides the trickling of water for untold ages has descended in times of rain, and for long periods after, into sandy basins at its feet. The dimensions of this vast slab are great, being over two miles long, over one mile through, and more than a quarter of a mile high. The great difference between it and Mount Olga is in the rock formation, for this is one solid granite stone, and is part and parcel of the original rock, which having been formed after its state of fusion in the beginning, has there remained, while Mount Olga has been thrown up subsequently from below. Mount Olga is the more wonderful and grotesque, Mount Ayers the more ancient and sublime. There is a permanent water here, but unlike the Mount Olga springs it is all in standing pools. There is excellent feeding-ground about this rock, though now the grass is all withered. This is certainly a most desirable and delectable spot for an explorer's depôt. The range farther still to the east seems farther away from here, than it did from Mount Olga. It is flat on the summit, and I suppose must be the same high and flat-topped mount I discovered from the Sentinel in August last.

Wednesday, 10th June.—Thermometer last night 54°, this morning 50°. We are encamped in the Roomy-Cave, and find it much warmer than outside. We rested the horses here to-day, as one of them is bad; and now, being about to undertake fresh discovery, albeit to the eastward, I wished to have the horses as fresh as possible at starting.

Thursday, 11th June.—Thermometer last night 58°, this morning 48°. We left Ayers-Rock and steered away to the east, towards the next range lying in that direction. We travelled twenty-three miles, all extemely heavy sandhills, with casuarina and triodia, and camped, of course, without water. Thermometer to-day, 64°.

Friday, 12th June.—Thermometer last night 50°, this morning 36°. We started early this morning, the day warmer than yesterday. Formby knocked up, and we had to leave him in the scrub. We travelled thirty miles to-day, and the flat-topped range is still seven or eight miles off, and looks most repulsive, as far as likelihoods for obtaining water are concerned. We camped again without water. Thermometer 74°. The country was now very scrubby and a perfect desert—worse travelling indeed, than Gibson's Desert itself.

Saturday, 13th June.—Thermometer last night 50°, this morning below zero. Mr. Tietkens and I rode over to the mount, because it was useless to take all the horses there, as now while we were near this mount I felt quite sure it was waterless, so we left them with Jimmy. We reached it in seven miles, the surrounding country being all scrub. At one and a-half miles we came to an outer escarpment of rocks, but between that and the mount more sandhills and thick scrub existed. We rode round this peculiar feature. It was many hundreds of feet high, and for about half its height its sides sloped, then the crown rested upon a perpendicular wall. It was almost circular and perfectly flat upon the top, having precisely the same vegetation and timber as grew upon the ground below. I don't know that it is accessible; I did not attempt to ascend it. To the north and about fifteen miles away was seen a portion of the not yet ended Amadeus Lake. To the east timbered ridges bounded the view. There was no water here whatever, though there were a few claypans. We were sixty miles from Ayers-Rock, and from all appearances we might have gone sixty or even a hundred miles farther, before we should reach any water. A large creek, called the "Lilla," joins the Finke-River upon this line of latitude, but where it arises no one knows. I thought to find some water here, and get on to it, and thence to the River Finke, but as it is, I shall return and make for that river by another line. We returned to Jimmy and the horses, and started the whole back for Ayers-Rock, and got back twenty-two miles upon our tracks the third night, without water. Thermometer to-day 74°.

Sunday, 14th June.—Thermometer last night 46°, this morning below zero. Horses strayed, and it was late before a start could be effected, as they had gone into the scrubs in several mobs. We had thirty-one miles to go to reach Ayers-Rock. The horses, well knowing where they were going, travelled well; but we did not reach the water till nine o'clock at night. On again reaching the spot where Formby had knocked up we searched for him and went a mile upon his tracks; but I could not afford the time. No doubt he had lain down and died not far off from where we had left him. Thermometer to-day 72°.

Monday, 15th June.—Thermometer last night 56°, this morning 44°. We must now kill another horse, as our supply is again done. I am sorry now, that I did not kill Formby before we left here, though he was scarcely fit even for explorers to eat. It appeared from the bones I found here that Mr. Gosse had killed one bullock, if not more; of course they must have strayed from the Telegraph-Line. I could not help wishing I could get one instead of killing a horse; however, I suppose it's the fortune of war. We commenced to erect a smoke-house to-day. Thermometer 74°.

Tuesday, 16th June.—Thermometer last night 62°, this morning 58°. Worked to-day at the smoke-house.

Wednesday, 17th June.—Thermometer last night 66°, this morning 54°; cloudy morning. We shifted the camp to the smoke-house, and everything being ready, down went poor Hollowback. He had been a very good horse. I bought him from Mr. Trew of the Globe. He was in what's called good working condition; but he had not a vestige of fat about him. The only adipose matter we could obtain from him was by boiling his bones, and the small quantity of oil thus obtained would only fry a few meals of steaks. When that was done, we used to fry them in water. The most favourite method of cooking the horseflesh was by first boiling, then pounding with the shoeing hammer, then cutting it up into small pieces, wetting the whole mass and making it into the form of a damper, with a pannikin of flour to seven or eight pounds of meat to bind it together; but the flour would not last, and those delicious horse-dampers were now things of the past.

Thursday, 18th June.—Thermometer last night 64°, this morning 66°; very windy morning. Thermometer 78°, cloudy day. Smoking Hollowback.

Friday, 19th June.—Thermometer last night 76°, this morning 58°. A warm night, a few sprinkles of rain. Some natives commenced yelling near the camp this morning. Three only made their appearance. They were the least offensive and most civil met by us in any of our travels. We gave them some of the bones and odd pieces of horse-meat, which seemed to give them great satisfaction, and they ate some pieces raw. They only remained about three hours. The day was windy and disagreeable. A blast of wind blew my only thermometer, which was hanging on a small sapling, and dashed it so violently against the ground, that it broke. One of the natives, who visited us to-day, had long curls waving in the wind, which hung down below his shoulders; the other two had chignons. Still smoking Hollowback. Mr. Tietkens had been using a small pair of bright steel plyers, when these natives were here. When the natives were gone, the plyers had also departed.

Saturday, 20th June.—Smoking horse and mending bags as usual.

Sunday, 21st June.—Smoking Hollowback.

Monday, 22nd June.—Smoking and packing up.

Tuesday, 23rd June.—I now intended to go to the north-north-eastward, to King's Creek, in the Gill's Range, of my former expedition, and which I presumed to be the spot, from whence Mr. Gosse had reached this rock; of course my Lake Amadeus lay in the road, but Mr. Gosse must have crossed it with his waggon, and I had no doubt his crossing-place would suit me also, though I did not expect to find any water between here and Gill's Range. We only went four miles this evening, and let the horses go back to the rock without hobbles.

Wednesday, 24th June.—Started at twelve o'clock. On the road we got water by digging at a native well, to which Mr. Gosse's dray-track took us. When we reached King's Creek we found it dry. At Penny's Creek, four miles east of it, there was plenty of water. We had now a fair and fertile track to the River Finke, discovered by me upon my former expedition, getting water at Petermann's Creek, fish and water at Middleton's Ponds; thence down the Palmer by Rogers's Pass and Briscoe's Pass, and on to the Finke, where there is a fine permanent waterhole at the junction.

Friday, 10th July.—Being at Crown-Point on the Finke-River, we met Mr. Alfred Frost and his teams going to the Alice-Springs. He was most hospitable and kind to us, and gave us plenty of white man's food, and we stayed a day with him and his party, and they gave us all the news of the world, and in return for their kindness we gave them all our remaining stock of Hollowback (for their dogs).

Monday, 13th July.—At about twelve o'clock we reached the Charlotte-Waters Telegraph-Station, and were most kindly received, clothed and fed by my excellent friend and namesake, Mr. Christopher Giles, jun., and all the staff at that establishment testified their pleasure at our return. Our welcome at the Peake was also most gratifying, by Mr. and Mrs. Blood, at the Telegraph-Station, and by Mr. John Bagot, at his cattle-station, who also supplied us with many necessaries.

* * * * * * * *

I have now but a few concluding remarks to make, for my second expedition is at an end, and those who have had the patience to wade through the verbose and rambling narrative of my wanderings are doubtless as willing to arrive at its conclusion as I was. I may truly say, that for twelve months I have been the well-wrought slave, not only of the sextant, the compass and the pen, but also of the shovel, the axe and the needle; there has been a continual strain upon brain and muscle. The leader of such an expedition as mine, with only three companions at starting and two at the finish, could not stand by and give orders for certain work to be performed, but must join in it and with the good example of heart and hand assist and cheer those, with whom he is associated. To my friend and second, Mr. Tietkens, I am under great obligations, for I have always found him (as my readers have doubtless perceived) ever ready and ever willing for the most arduous and disagreeable of our many undertakings. My expedition has in its main objects been unsuccessful, and my most sanguine hopes have been destroyed; as for years I had cherished the thought of being the first to cross this Continent from east to west, and it may have been the hope of others also. I know, at starting, a great deal was expected from me; and if I have not fulfilled the hopes of my friends, I can only console them with the fact, that I could not even fulfil my own; but if it is conceded, that I have done my devoir as an Australian explorer should, then I shall be satisfied. Many trials and many bitter hours must the explorer of such a region experience, whether successful or unsuccessful in his attempt. The life of a man is to be held at no more than a moment's purchase, while engaged in such an enterprise; and it may be truly said, we have passed through a baptism worse indeed than that of fire—the baptism of *no water*. Three expeditions, starting almost simultanously from different points, to penetrate to the west, was a remarkable point in the history of the exploration of this Continent, and I with my diminutive force and sparse equipment had scarcely a right to contend in the same field, while the other expeditions were furnished with camels. It is with no feelings of envy whatever, that I hail my old friend Colonel Warburton as the victor in this contest, in which he also nearly fell upon the field of his renown. His gallant old hand has planted the flag, and his gallant old brows shall wear the crown—the laurel-crown of triumph and of victory, for peace hath her victories no less renowned than war; and it will surprise no one, if our gracious Sovereign, with her honour-giving hand, should lay the glittering accolade of knighthood upon his shoulder; and, in such a case, the liberal and enlightened patron and patriot, at whose desire and support this great undertaking was accomplished, will not be forgotten. I, the defeated and unsuccessful, can lay no flattering unction to my soul, and having abandoned the task, shall return again into that seclusion and obscurity, unknowing and unknown, from whence I may for a brief moment meteor-like have emerged. That I shall ever again take the field is more than I can undertake to say. But—

> " . . . . . Yet the charmed spell
> Which summons man to high discovery
> Is ever vocal in the outward world,
> Though they alone may hear it who have hearts
> Responsive to its tone. . . . . . ."

I will now only add, that a line of permanent waters has been found to

Sladen-Water and Fort McKellar, on the Rawlinson-Range, which latter cannot be further away from the Murchison-River than 350 or 400 miles; out of that distance, or rather into it, I penetrated 120 miles, and saw ranges lying about thirty miles further away; these I have named the Alfred and Marie Ranges, after their Royal Highnesses the Duke and Duchess of Edinburgh. At what price these ranges were seen, I need not now repeat, but they are there, and it is highly probable, that water exists in them also. From these ranges, only one water between would carry the traveller to the River Murchison; but it is only with camels that there is much likelihood of a successful and permanently valuable issue. In case of any future attempt only one gentleman in the whole of Australia can supply the means of its accomplishment, and to him the country at large must be in future, as it is at present, indebted for ultimate discoveries. Of course that gentleman is the Hon. Thomas Elder.

I have now to thank those gentlemen in Victoria, who so kindly subscribed to the fund for my expedition, and to the gentlemen of the late Ministry of South Australia, who subsidised the fund; and lastly, my most excellent friend, Baron Mueller, who raised the fund on my behalf, and generously brought me before the public as a man, suitable to undertake the exploration of the unknown portions of this Continent; and if the world's opinion will coincide with his upon the matter, I at least shall be satisfied.

ERNEST GILES.

ALONE IN THE DESERT.

PLANTS

COLLECTED BY MR. GILES,

DURING HIS

GEOGRAPHIC EXPLORATION OF CENTRAL AUSTRALIA

IN 1872, 1873 AND 1874.

EXAMINED BY

BARON FERD. VON MUELLER,

C.M.G., M. & PH. D., F.R.S.

LIST OF PLANTS.

Dilleniaceæ.

Hibbertia glaberrima, F. M., Fragm. III., 1.
Mount Olga, Glen of Palms.

Cruciferæ.

Menkea sphærocarpa, F. M., Fragm. VIII., 223.
Near Mount Olga.
Lepidium papillosum, F. M. in Linnæa XXV., 370.
Between the Alberga and Mount Olga.
Lepidium phlebopetalum, F. M., Plants of Vict. I., 47.
Between the River Finke and Lake Eyre.
Sisymbrium trisectum, F. M., Transact. Vict. Inst. I., 114.
Near Lake Eyre and Mount Olga.

Capparideæ.

Cleome viscosa, L. Sp. Pl., 938.
Rawlinson's Range.
Capparis Mitchelli, Lindl. in Mitch. Three Exped. I., 315.
MacDonnell's Range, Mount Udor.

Pittosporeæ.

Pittosporum phillyroides, Cand. Prodr. I., 347.
Between the Alberga and Mount Olga, also on Gosse's Range.

Droseraceæ.

Drosera Indica, L. Sp., 403.
Rawlinson's Range.
Drosera Burmanni, Vahl. Symb. III., 50.
MacDonnell's Range.

Polygaleæ.

Comesperma silvestre, Lindl. in Mitch. Trop. Austr., 342
Between MacDonnell's and Gill's Ranges.

Violaceæ.

Ionidium aurantiacum, F. M. in Benth. Fl. Austr. I., 102.
MacDonnell's Range.

Geraniaceæ.

Oxalis corniculata, L. Sp., 624.
Between the Alberga and Mount Olga.

Tiliaceæ.

Corchorus sidoides, F. M., Fragm. III., 9.
MacDonnell's Range.

Malvaceæ.

Hibiscus Farragei, F. M., Fragm. VIII., 241.
MacDonnell's Range. Messrs. Forrest found the same species on their 78th camp.

Hibiscus Sturtii, Hook. in Mitch. Trop. Austr., 363.
Rawlinson's Range.

Hibiscus microchlaenus, F. M., Fragm. II., 116.
Rawlinson's Range.

Gossypium Sturtii, F. M., Fragm. III., 6.
On Mount Olga, also towards the Alberga, Gosse's Range, and MacDonnell's Range. Found likewise by Messrs. Forrest on their 78th camp.

Abutilon diplotrichum, F. M. in Linnæa XXV., 380.
Between Lake Eyre and the River Finke.

Abutilon halophilum, F. M. in Linnæa XXV., 381.
Near Lake Eyre.

Sida cardiophylla, F. M., Fragm. VIII., 242.
Rawlinson's Range.

Sida inclusa, Benth., Flor. Austr. I., 197.
Rawlinson's Range, MacDonnell's Range.

Sida cryphiopetala, F. M., Fragm. II., 4.
MacDonnell's Range.

Sida virgata, Hook. in Mitch. Trop. Austr., 361.
Mount Olga.

Sida petrophila, F. M. in Linnæa XXV., 381.
MacDonnell's Range.

Sida corrugata, Lindl. in Mitch. Three Exped. II., 13.
Lake Eyre, Mount Olga, Gosse's Range, MacDonnell's Range, Lake Amadeus.

Malvastrum spicatum, As. Gr. Plant Fendl., 23.
Near Lake Eyre.

Plagianthus glomeratus, Benth. in Journ. of Linn. Soc. VI., 103.
Near Lake Eyre.

Sterculiaceæ.

Keraudrenia nephrosperma, Benth., Fl. Austr. I., 246.
Mount Olga, MacDonnell's Range.

Keraudrenia Hookeriana, Walp., Annal. II., 164.
MacDonnell's Range.

Rulingia magniflora, F. M., Fragm. VIII., 223.
Mount Olga.

Rulingia loxophylla, F. M., Fragm. I., 68.
MacDonnell's Range.

Brachychiton Gregorii, F. M. in Hook. Kew Mis. IX., 199.
Mount Stevenson, MacDonnell's Range, Carmichael's Creek, Mount Udor; also near Messrs. Forrest's camp of the 20th May. The specific position, in the absence of flowers and fruit, not to be ascertained beyond doubts from the material secured.

Frankeniaceæ.

Frankenia pauciflora, Cand. Prodr. I., 350.
Lake Eyre, River Finke.

ZYGOPHYLLEÆ.

Tribulus terrestris, L. Sp., 554.
Rawlinson's Range.

Tribulus Hystrix, R. Br., App. to Sturt's Centr. Austr., 6.
Near Lake Amadeus.

Zygophyllum fruticulosum, Cand. Prodr. I., 705.
Near Lake Eyre.

SAPINDACEÆ.

Atalaya hemiglauca, F. M. in Benth. Fl. Austr. I., 463.
MacDonnell's Range and Lake Amadeus.

Dodonæa viscosa, L. Mantiss., 238.
Alberga, Mount Olga, Rawlinson's Range, Barrow's Range. The allied *D. petiolaris* was met by Messrs. Forrest near their camps of 20th May and 12th and 15th July. *D. microzyga*, F. M., Plants of Stuart's Exped., 1862, p. 12, is known from the Neale River.

Diplopeltis Stuartii, F. M., Fragm. III., 12.
MacDonnell's Range.

PHYTOLACCEÆ.

Codonocarpus cotinifolius, F. M., Plants of Vict. I., 200.
Between the Alberga and Mount Olga; also near Forrest's track of the 15th July, on Mount Hale and Hoslup's Well.

Gyrostemon ramulosus, Desf. in Mem. Du Mus. VI., 17 t. 6.
Glen of Palms.

Cyclotheca Australasica, Mog. in Cand. Prodr. XIII., Sect. II., 38.
Mount Olga, Rawlinson's Range, Barrow's Range.

CARYOPHYLLEÆ.

Polycarpæa corymbosa, Lam. Ill. N., 2798.
Glen of Palms.

FICOIDEÆ.

Trianthema crystallina, Vahl., Symb. I., 32.
Near Lake Eyre.

Aizoon zygophylloides, F. M., Fragm. VII., 129.
Between Lake Eyre and the River Finke.

PORTULACEÆ.

Calandrinia Balonnensis, Lindl. in Mitch. Trop. Austr., 148.
MacDonnell's Range.

Portulaca oleracea, L. Sp. Pl., 638.
Towards MacDonnell's Range.

SALSOLACEÆ.

Rhagodia nutans, R. Br., Prodr., 408.
Lake Eyre.

Rhagodia spinescens, R. Br., Prodr., 408.
Lake Eyre.

Chenopodium carinatum, R. Br., Prodr., 407.
Between the Alberga and Mount Olga.

Babbagia dipterocarpa, F. M., Rep. on Babb. Pl., 21.
Lake Eyre.

Kochia villosa, Lindl. in Mitch. Trop. Austr., 91.
Between the Alberga and Mount Olga.

AMARANTACEÆ.

Hemichroa mesembryanthema, F. M., Fragm. VIII., 38.
Lake Eyre.
Euxolus Mitchelli, Amarantus Mitchelli, Benth., Fl. Austr. V., 214.
Lake Eyre.
Alternanthera nodiflora, R. Br., Prodr., 417.
MacDonnell's Range.
Ptilotus obovatus, F. M., Fragm. VI., 228.
Between the Alberga and Mount Olga; MacDonnell's and Rawlinson's Ranges.
Ptilotus alopecuroides, F. M., Fragm. VI., 227.
Between the Alberga and Mount Olga.
Ptilotus nobilis, F. M., Fragm. VI., 227.
Mount Olga.
Ptilotus Hoodii, F. M., Fragm. VIII., 232.
Mount Olga.
Ptilotus helipteroides, F. M., Fragm. VI., 231.
Between the Alberga and Mount Olga; also Barrow's Range.
Ptilotus hemisteirus, F. M., Fragm. VI., 231.
Between the Alberga and Mount Olga.

NYCTAGINEÆ.

Boerhaavia repanda, Willd., Sp. Pl. I., 22.
Lake Eyre.
Boerhaavia diffusa, L. Sp. Pl., 4.
Lake Amadeus.

LEGUMINOSÆ.

Daviesia arthropoda, F. M., Fragm. VIII., 225.
Mount Olga.
Brachysema Chambersii, F. M. in Benth. Fl. Austr. II., 13.
Mount Olga; MacDonnell's Range.
Isotropis atropurpurea, F. M., Fragm. III., 16.
Mount Olga.
Burtonia polyzyga, Benth., Fl. Austr. II., 51.
MacDonnell's Range.
Mirbelia oxyclada, F. M., Fragm. IV., 12.
MacDonnell's and Rawlinson's Ranges.
Gastrolobium grandiflorum, F. M., Fragm. III., 17.
Glen of Palms.
Psoralea patens, Lindl. in Mitch. Three Exped. II., 9.
Between Lake Eyre and Mount Olga. *P. balsamica* is known from MacDonnell's Range.
Crotalaria Cunninghami, R. Br., App. to Sturt's Exped., 8.
Rawlinson's Range; also on Messrs. Forrest's 71st camp.
Crotalaria dissitiflora, Benth. in Mitch. Trop. Austr., 386.
Lake Eyre.
Clianthus Dampierii, A. Cunn. in Trans. Hort. Soc. Lond., Sec. Ser. I., 522.
Mount Whitby.
Swainsona phacoides, Benth. in Mitch. Trop. Aust., 363.
MacDonnell's Range.
Swainsona unifoliolata, F. M., Fragm. VIII., 226.
Between the Alberga and Mount Olga; also on Rawlinson's Range. Several other species of *Swainsona*, but in an imperfect state, occur in the collection, also a species of *Tephrosia*.

Lotus Australis, Andr., Bot. Reg., t. 624.
Lake Eyre.

Caulinia prorepens, F. M., Fragm. VIII., 225.
Between the Alberga and Mount Olga; seen also by Messrs. Forrest near their camps of the 13th and 15th July.

Indigofera monophylla, Cand. Prodr. II., 222.
MacDonnell's Range.

Indigofera brevidens, Benth. in Mitch. Trop. Austr., 385.
Between Lake Eyre and the River Finke; also Glen of Palms, MacDonnell's Range, Rawlinson's Range, between Mount Olga and Barrow's Range. (*I. villosa* is also known from MacDonnell's Range.)

Erythrina Vespertilio, Benth. in Mitch. Trop. Austr., 218.
MacDonnell's Range, Mount Udor.

Bauhinia Leichhardtii, F. M. in Transact. Vict. Inst. III., 50.
Occurs also in many of the central regions of the continent.

Cassia notabilis, F. M., Fragm. III., 28.
Mount Olga, Rawlinson's Range.

Cassia venusta, F. M., Fragm. I., 165.
MacDonnell's Range.

Cassia pleurocarpa, F. M., Fragm. I., 223.
Between Lake Eyre and the River Finke; also between the Alberga and Mount Olga, MacDonnell's Range.

Cassia desolata, F. M. in Linnæa XXV., 389.
Mount Olga, Rawlinson's Range.

Cassia artemisioides, Gaud. in Cand. Prodr. II., 495.
From the Alberga to Mount Olga and Barrow's Range.

Petalostylis labicheoides, R. Br., App. to Sturt's Centr. Austr., 17.
Glen of Palms; between the Alberga and Mount Olga, and towards Barrow's Range.

Acacia Sentis, F. M. in Journ. Linn. Soc. III., 128.
Between Mount Olga and Barrow's Range.

Acacia patens, F. M. in Journ. Linn. Soc. III., 120.
Mount Olga and MacDonnell's Range.

Acacia spondylophylla, F. M., Fragm. VIII., 243.
Glen of Palms; MacDonnell's and Rawlinson's Ranges.

Acacia lycopodifolia, A. Cunn. in Hook. Icon., 172.
MacDonnell's Range.

Acacia minutifolia, F. M., Fragm. VIII., 243.
Mount Olga.

Acacia strongylophylla, F. M., Fragm. VIII., 226.
Between the Alberga and Mount Olga, Glen of Palms, MacDonnell's Range.

Acacia salicina, Lindl. in Mitch. Three Exped. II., 20.
Between the Alberga and Mount Olga. MacDonnell's Range; also towards Lake Amadeus and Barrow's Range.

Acacia aneura, F. M. in Linnæa XXVI., 627.
Between Mount Olga and Barrow's Range.

Numerous other species of *Acacia* were gathered, but not found in flower or fruit, hence are not with certainty referable to the respective species of this great genus.

EUPHORBIACEÆ.

Adriana tomentosa, Gaud. in Ann. Sc. Nat., Prem. Ser. VI., 223.
From the Alberga to Mount Olga. MacDonnell's Range, Barrow's Range.

Euphorbia Drummondi, Boiss., Cent. Euph., 14.
Finke's River.

Euphorbia eremophila, A. Cunn. in Mitch. Austr., 348.
Lake Eyre; MacDonnell's Range.

URTICEÆ.

Ficus platypoda, A. Cunn. in Hook. Lond. Journ. VI., 561.
Between the Alberga and Mount Olga, Ayers' Range, Gill's Range; found also by Messrs. Forrest near their 70th and 78th camps.

Ficus orbicularis, A. Cunn. in Hôok. Lond. Journ. VII., 426.
Glen of Palms.

Parietaria debilis, G. Forst., Prodr., 73.
Mount Olga.

RHAMNACEÆ.

Spyridium spathulatum, F. M. in Benth. Fl. Austr. I., 430.
Glen of Palms.

MYRTACEÆ.

Calycothrix longiflora, F. M., Fragm. I., 12.
Between the Alberga and Mount Olga; MacDonnell's Range.

Thryptomene Maisonneuvii, F. M., Fragm. IV., 64.
On Mount Olga, also towards the Alberga.

Thryptomene flaviflora, F. M., Fragm. VIII., 13.
MacDonnell's Range.

Baeckea polystemonea, F. M., Fragm. II., 124.
MacDonnell's Range.

Eucalyptus pachyphylla, F. M. in Journ. Linn. Soc. III., 98.
Glen of Palms.

STACKHOUSIACEÆ.

Macgregoria racemigera, F. M., in Caruel's Giorn., 1873, p. 129.
MacDonnell's Range; between Mount Olga and Barrow's Range.

Stackhousia megaloptera, F. M., Fragm. VIII., 35.
MacDonnell's Range.

CUCURBITACEÆ.

Mukia scabrella, Arn. in Hook. Journ. III., 276.
Rawlinson's Range.

Cucumis trigonus, Roxb., Flor. Indic. III., 722.
MacDonnell's Range.

LORANTHACEÆ.

Loranthus Exocarpi, Behr in Linn. XX., 624.
Musgrave Range. Observed also at Forrest's camps of 23rd June and 15th July.

SANTALACEÆ.

Santalum lanceolatum, R. Br., Prodr., 256.
Mount Olga, Rawlinson's Range, Lake Amadeus.

Santalum acuminatum, A. de Cand. Prodr. XIV., 684.
Mount Olga, MacDonnell's Range, Mount Udor, Lake Amadeus, Musgrave Range, Fort Mueller, Petermann's Range.

Anthobolus exocarpoides, F. M., Fragm. IX., ined.
MacDonnell's Range.

PROTEACEÆ.

Hakea multilineata, Meissn. in Lehm. Pl. Preiss. II., 261.
Between the Alberga and Mount Olga.

Hakea lorea, R. Br., Prot. Nov., 25.
Glen of Palms, MacDonnell's, Petermann's, and Rawlinson's Ranges. Noted by Messrs. Forrest, as occurring near their 70th camp.

Grevillea stenobotrya, F. M., Fragm. IX., ined.
MacDonnell's Range; also at Forrest's 71st camp.

Grevillea juncifolia, Hook. in Mitch. Trop. Austr., 341.
Glen of Palms, MacDonnell's Range, Mount Olga, and towards the Alberga.

Grevillea pterosperma, F. M. in Trans. Phil. Soc. Vict. I., 22.
Mount Olga.

Grevillea Wickhami, Meissn. in Cand. Prodr. XIV., 380.
Glen of Palms, Gosse's Range, MacDonnell's Range; towards Lake Amadeus.

THYMELEÆ.

Pimelea trichostachya, Lindl. in Mitch. Trop. Austr., 355.
Between the Alberga and Mount Olga. Gosse's Range.

Pimelea ammocharis, F. M. in Hook. Kew Misc. IX., 24.
Between Mount Olga and Barrow's Range.

UMBELLIFERÆ.

Didiscus glaucifolius, F. M. in Linnæa XXV., 395.
Between the Alberga and Mount Olga; gathered also on 28th June by Messrs. Forrest, on their track.

Hydrocotyle trachycarpa, F. M. in Linnæa XXV., 394.
Between the Alberga and Mount Olga.

RUBIACEÆ.

Pomax umbellata, Soland. in Gaertn. Fruct. I., 112.
MacDonnell's Range.

Plectronia latifolia, Benth. et Hook. Gen. Pl. II., 110.
MacDonnell's Range.

COMPOSITÆ.

Aster subspicatus, F. M., Fragm. V., 68.
MacDonnell's Range.

Aster megalodontus, F. M., Fragm. VIII., ined.
Mount Olga.

Aster Ferresii, F. M., Fragm. V., 75.
MacDonnell's Range.

Calotis lappulacea, Benth. in Hueg. Enum., 60.
Between the Alberga and Mount Olga.

Pluchea Eyrea, F. M., Rep. on Babb. Pl., 11.
Mount Olga, Macdonnell's Range.

Minuria leptophylla, Cand. Prodr. V., 298.
Between Lake Eyre and the River Finke, thence to Mount Olga and Lake Amadeus.

Flaveria Australasica, Hook. in Mitch. Trop. Austr., 118.
Lake Eyre.

Gnephosis codonopappa, F. M., Fragm. IX., ined.
Beyond Lake Eyre.

Angianthus tomentosus, Wendl. Coll. II., 31 t. 48.
Between Mount Olga and Barrow's Range.

Calocephalus platycephalus, Benth., Fl. Austr. III., 576.
MacDonnell's Range.

Myriocephalus Stuartii, Benth., Fl. Austr. III., 560.
Between the Alberga and Mount Olga.

Pterocaulon sphacelatus, Benth. et Hook., Gen. Pl. II., 295.
Between the Alberga and Mount Olga, also on Rawlinson's Range.

Ixiolaena tomentosa, Sond. et Muell. in Linnaea XXV., 504.
Lake Eyre.
Helichrysum Thomsoni, F. M., Fragm. VIII., 45.
MacDonnell's Range. Mount Olga.
Helichrysum Ayersii, F. M., Fragm. VIII., 167.
Mount Olga.
Helichrysum semifertile, F. M., Rep. on Babb. Plants, p. 14.
Between the Alberga and Mount Olga.
Helichrysum Davenporti, F. M., Fragm. III., 32.
Between the Alberga and Mount Olga.
Helichrysum Cassinianum, Gaud. in Freyc. Voy. Bot., 466, t. 87.
MacDonnell's Range; also between the Alberga and Mount Olga.
Helichrysum lucidum, Henck. Adumb. Ann., 1806.
Between the Alberga and Mount Olga, Glen of Palms, Rawlinson's Range.
Helichrysum apiculatum, Cand. Prodr. VI., 195.
Rawlinson's Range.
Helichrysum rutidolepis, Cand. Prodr. VI., 194.
Between the Alberga and Mount Olga.
Helipterum floribundum, Cand. Prodr. VI., 217.
Between the Alberga and Mount Olga.
Helipterum Tietkensii, F. M., Fragm. VIII., 227.
Between the Alberga and Mount Olga.
Helipterum incanum, Cand. Prodr. VI., 215.
Between the Alberga and Mount Olga.
Helipterum stipitatum, F. M. in Benth. Fl. Austr. III., 643.
MacDonnell's Range.
Helipterum Charsleyæ, F. M., Fragm. VIII., 168.
Lake Amadeus.
Gnaphalium luteo-album, L. Sp. Pl., 1196.
Mount Olga.
Gnaphalium Japonicum, Thunb., Fl. Jap., 311.
Mount Olga.
Senecio Gregorii, F. M. in Greg. Rep. on Leich. Search, p. 7.
Between the Alberga and Mount Olga, MacDonnell's Range.
Senecio lautus, G. Forst., Prodr., 91.
Between the Alberga and Mount Olga.
Senecio magnificus, F. M. in Linnæa XXV., 418.
Between the Alberga and Mount Olga.
Erechtites picridioides, Turcz. in Bull. de Mosc., 1851, part I., 200.
Mount Olga.
Sonchus oleraceus, Linné, Sp. Pl., 1116.
Mr. Giles records this in his Journal as abundant on the banks of the Finke-River, towards its source.

CAMPANULACEÆ.

Wahlenbergia gracilis, A. de Cand. Monogr. des Camp., 142.
Mount Olga, Barrow's Range, Lake Amadeus.
Lobelia heterophylla, Labill. Specim. I., 52, t. 74.
Between Mount Olga and Barrow's Range.
Isotoma petræa, F. M. in Linnæa XXV., 420.
Between the Alberga and Mount Olga, MacDonnell's Range.

GOODENOVIACEÆ.

Brunonia Australis, Sm. in Transact. Linn. Soc. X., 367, t. 28.
Between the Alberga and Mount Olga, MacDonnell's Range.

Goodenia Vilmoriniæ, F. M., Fragm. III., 19, t. 16.
Mount Olga.
Goodenia heterochila, F. M., Fragm. III., 142.
Between Mount Olga and Barrow's Range.
Goodenia Mueckeana, F. M., Fragm. VIII., 56.
Between Mount Udor and Gill's Range; also on or near Mount Olga.
Goodenia Ramelii, F. M., Fragm. III., 20, t. 17.
Between the Alberga and Mount Olga; also on Rawlinson's Range and towards Barrow's Range.
Leschenaultia divaricata, F. M., Fragm. III., 33.
Lake Amadeus.
Leschenaultia striata, F. M., Fragm. VIII., 245.
Mount Olga.
Catosperma Muelleri, Benth., Fl. Austr. IV., 83.
Between the Alberga and Mount Olga.
Scaevola collaris, F. M., Rep. on Babb. Plants, 15.
Lake Eyre.
Scaevola spinescens, R. Br., Prodr., 568.
Lake Eyre.
Scaevola depauperata, R. Br., Append. to Sturt's Centr. Austr., 20.
MacDonnell's Range.
Velleya connata, F. M. in Hook. Kew Misc. VIII., 162.
MacDonnell's Range.

STYLIDEÆ.

Stylidium floribundum, R. Br., Prodr., 569.
MacDonnell's Range.

ASPERIFOLIÆ.

Heliotropium asperrimum, R. Br., Prodr., 493.
Between the Alberga and Mount Olga, MacDonnell's Range.
Heliotropium undulatum, Vahl., Sym. I., 13.
Near Lake Eyre. Messrs. Forrest brought this species from Hoslup Well.
Cynoglossum Drummondi, Benth., Fl. Austr. IV., 409.
On Mount Olga and towards the Alberga.
Trichodesma Zeilanicum, R. Br., Prodr., 496.
From the Alberga to Mount Olga and MacDonnell's Range.
Halgania anagalloides, Endl. in Ann. des Wien. Mus. II., 204.
MacDonnell's Range.
Halgania cyanea, Lindl. Bot. Reg. XXV., App., 40.
MacDonnell's and Petermann's Ranges.

LABIATÆ.

Plectranthus parviflorus, Henck. Adumb., 1806.
Between Mount Olga and Barrow's Range.
Microcorys Macredieana, F. M., Fragm. VIII., 231.
Rawlinson's Range.
Prostanthera striatiflora, F. M. in Linnæa XXV., 425.
From the Alberga to Mount Olga; also on Gosse's Range and MacDonnell's Range.
Prostanthera Wilkieana, F. M., Fragm. VIII., 230.
Between Mount Olga and Barrow's Range.
Teucrium racemosum, R. Br., Prodr., 504.
Lake Eyre, Lake Amadeus, Finke River.

VERBENACEÆ.

Newcastlia bracteosa, F. M., Fragm. VIII., 49.
MacDonnell's Range; between Mount Olga and Warburton's Range; Gill's Range.

Newcastlia cephalantha, F. M., Fragm. IX., ined.
Between the Alberga and Mount Olga.

Newcastlia spodiotricha, F. M., Fragm. III., 21, t. 21.
MacDonnell's and Rawlinson's Ranges.

Dicrastylis Dorani, F. M., Fragm. VIII., 230.
Rawlinson's Range.

Dicrastylis ochrotricha, F. M., Fragm. IV., 161.
Between Mount Olga and Barrow's Range, occurring also in Messrs. Forrest's collection from Mount Moore and Alexander's Springs.

Dicrastylis Beveridgei, F. M., Fragm. VIII., 50.
Between Mount Udor and Gill's Range, also on Mount Olga.

Dicrastylis Gilesii, F. M., Fragm. VIII., 229.
Between the Alberga and Mount Olga; Glen of Palms.

Chloanthes Lewellini, F. M., Fragm. VIII., 50.
Mount Olga; MacDonnell's Range.

MYOPORINÆ.

Eremophila Macdonnelli, F. M., Rep. on Babb. Plants, 18.
Between Lake Eyre and the River Finke.

Eremophila Willsii, F. M., Fragm, III., 21, t. 20.
Between the Alberga and Mount Olga; Rawlinson's Range.

Eremophila Berryi, F. M., Fragm. VIII., 228.
Musgrave Range (Gosse).

Eremophila Goodwini, F. M., Rep. on Babb. Plants, 17.
Beyond Lake Eyre, Glen of Palms, MacDonnell's Range.

Eremophila maculata, F. M. in Papers of the Roy. Soc. of Tasm. III., 297.
Lake Eyre.

Eremophila Brownii, F. M. in Papers of the Roy. Soc. of Tasm. III., 297.
MacDonnell's Range.

Eremophila Sturtii, R. Br., App. to Sturt's Centr. Austr., 85.
MacDonnell's Range.

Eremophila Gilesii, F. M., Fragm. VIII., 49.
MacDonnell's Range.

Eremophila longifolia, F. M. in Papers of the Roy. Soc. of Tasm. III., 295.
Gosse's Range; MacDonnell's Range.

Eremophila latifolia, F. M. in Linnæa XXV., 428.
Between the Alberga and Mount Olga.

Eremophila alternifolia, R. Br., Prodr., 518.
Mount Olga.

Eremophila Latrobei, F. M. in Papers of the Roy. Soc. of Tasm. III., 294.
Mount Olga; Rawlinson's Range; MacDonnell's Range.

Eremophila Elderi, F. M., Fragm. VIII., 228.
Rawlinson's Range.

Eremophila Hughesii, F. M., Fragm. VIII., 228.
Rawlinson's Range.

Eremophila Gibsoni, F. M., Fragm. VIII., 227.
Between Mount Olga and the Alberga.

Eremophila scoparia, F. M. in Papers of the Roy. Soc. of Tasm. III., 296.
About Lake Eyre.

Myoporum Cunninghami, Benth. in Hueg. Enum., 78.
Glen of Palms.

JASMINEÆ.

Jasminum lineare, R. Br., Prodr., 521.
MacDonnell's Range; Gosse's Range.
Jasminum calcareum, F. M., Fragm. I., 212.
MacDonnell's Range.

CONVOLVULACEÆ.

Convolvulus erubescens, Sims, Bot. Mag., t. 1067.
MacDonnell's Range.
Evolvulus linifolius, L. Sp. Pl., 392.
MacDonnell's Range.
Breweria rosea, F. M., Fragm. I., 233.
Between the Alberga and Mount Olga, Glen of Palms, MacDonnell's Range.

BIGNONIACEÆ.

Tecoma Australis, R. Br., Prodr., 471.
Mount Olga, Rawlinson's Range.

ASCLEPIADEÆ.

Sarcostemma Australe, R. Br., Prodr., 463.
Rawlinson's Range.
Marsdenia Leichhardtiana, F. M., Fragm. V., 160.
MacDonnell's Range.

ACANTHACEÆ.

Justicia procumbens, L. Fl. Zeil., 19.
Mount Olga and towards Lake Eyre.

GENTIANEÆ.

Erythraea Australis, R. Br., Prodr., 451.
Between Mount Olga and Barrow's Range, MacDonnell's Range.

SCROPHULARINÆ.

Mimulus gracilis, R. Br., Prodr., 439.
Rawlinson's Range.
Stemodia viscosa, Roxb., Pl. Coromand. II., 33, t. 163.
Rawlinson's Range.
Stemodia pedicellaris, F. M., Fragm. VIII., 231.
Rawlinson's Range.

SOLANACEÆ.

Anthrotroche Blackii, F. M., Fragm. VIII., 232.
Between Mount Olga and Barrow's Range.
Anthocercis Hopwoodii, F. M., Frag. II., 138.
Near Mount Liebig.
Nicotiana suaveolens, Lehm., Hist. Nicot., 43.
Between the Alberga and Mount Olga; Glen of Palms; Lake Amadeus.
Solanum esuriale, Lindl. in Mitch. Three Exped. II., 43.
Lake Eyre; thence to MacDonnell's Range.

Solanum ferocissimum, Lindl. in Mitch. Three Exped. II., 58.
MacDonnell's Range.

Solanum ellipticum, R. Br., Prodr., 446.
Between the Alberga and Mount Olga; thence to Barrow's Range MacDonnell's Range.

Solanum petrophilum, F. M. in Linnæa XXV., 433.
Mount Olga.

Solanum lacunarium, F. M. in Trans. Phil. Soc. Vict. I., 18.
Lake Eyre.

Datura Leichhardti, F. M. in Trans. Phil. Soc. Vict. I., 20.
Between the River Finke and the Glen of Palms.

PRIMULACEÆ.

Samolus repens, Pers. Synops. I., 171.
Between Mount Olga and Barrow's Range.

CASUARINEÆ.

Casuarina Decaisneana, F. M., Fragm. I., 61.
From the Alberga and Finke River to Mount Olga; Gardiner's and MacDonnell's Ranges; Glen of Palms; also near Musgrave's Range, and on Rawlinson's, Petermann's, and Barrow's Ranges; Gibson's Desert.

CYCADEÆ.

Encephalartos Macdonnelli, F. M. in Vers. Akad. Wet. Amsterdam, XV., 376.
On Neale's River, found by J. M. Stuart, and probably the same species on Gill's Range.

CONIFERÆ.

Callitris verrucosa, R. Br. in Memoir. du Mus. Paris XIII., 74.
It is supposed that it is this species, which was seen on the River Finke, Lake Amadeus, and in the MacDonnell's, Gill's, Rampart's, Musgrave's, and Gosse's Ranges, as it is the only one hitherto recorded from Central Australian collections.

LILIACEÆ.

Thysanotus sparteus, R. Br., Prodr., 283.
Between Mount Olga and Barrow's Range.

Anguillaria Australis, F. M., Fragm. VII., 74.
Between Lake Eyre and the River Finke. A species of Xanthorrhœa, reaching a height of twelve feet, was seen on the ranges along Rudall's Creek, but no specimen for examination was secured.

PALMÆ.

Livistona Mariæ, F. M., Fragm. IX., ined.
Glen of Palms. Height up to 60 feet.

TYPHACEÆ.

Typha Muelleri, Rohrb. in Verhandl. Brandenb., 1869, p. 95.
It is probably this species which is recorded in the Journal as occurring in the swamps of Rawlinson's Range.

Gramineæ.

Andropogon laniger, Desf., Fl. Atlant. II., 379.
Between the Alberga and Mount Olga.

Eriachne scleranthoides, F. M., Fragm. VIII., 233.
Mount Olga.

Pappophorum commune, F. M. in Greg. Rep. on Leichh. Search, App., p. 10.
MacDonnell's Range.

Panicum Pseudo-Neurachne, F. M., Fragm. VIII., 199.
Lake Amadeus.

Eleusine cruciata, Lam. Encyc., t. 48, f. 2.
Lake Eyre; between the Alberga and Mount Olga.

Aristida stipoides, R. Br., Prodr., 174.
Between the Alberga and Mount Olga.

Bromus arenarius, Labill., Specim. I., 23, t. 28.
Between the Alberga and Mount Olga.

Festuca irritans, F. M., Chath. Isl. Veget., 59 (*Triodia irritans*, R. Br. Pr., 182).
Dispersed widely through the deserts, and called *Spinifex* by the explorers.

Cyperaceæ.

Cyperus textilis, Thunb., Prodr. Pl. Cap., 18.
MacDonnell's Range.

Filices.

Cheilanthes tenuifolia, Swartz, Syn. Fil., 129.
Rawlinson's Range; between the Alberga and Mount Olga.

Cheilanthes vellea, F. M., Fragm. V., 123.
Between the Alberga and Mount Olga; also on MacDonnell's Range. *C. Reynoldsii*, discovered by Mr. Gosse, does not occur in Mr. Giles's collection, and is probably very local.

Mr. Giles's collection contains also species of the genera *Vigna, Tephrosia, Melaleuca, Callistemon, Haloragis, Pterigeron, Brachycome, Dampiera, Ipomœa, Morgania, Enchylæna*, and *Atriplex;* as also additional species of *Rulingia, Abutilon, Sida, Dodonæa, Euphorbia, Spyridium, Acacia* (many), *Eucalyptus, Scævola, Goodenia, Eremophila, Heliotropium, Rhagodia, Ptilotus, Hakea*, and *Panicum;* but none in a state sufficiently advanced to admit of ascertaining their precise specific position.

A preliminary enumeration of the plants of Mr. Giles's First Expedition appeared in Caruel's *Nuovo Giornale Botanico Italiano, 1873*, pp. 127-129. Additional localities of many Central-Australian plants from John MacDouall Stuart's collections are recorded in the Official Report on the Melbourne Botanic Department for 1862, pp. 11-15.

MELBOURNE:
M'CARRON, BIRD AND CO., PRINTERS,
37 FLINDERS LANE WEST.